Acclaim for *Much Ado About Numbers*

"You can count on both Rob Eastaway and Shakespeare to make mathematics entertaining, never more so than in this elegant tome which will make your Pericles tingle as you like it."
—Sir Tim Rice

"A spectacular journey.... Highly recommended!"
—Dr. Simon Smith, associate professor of Shakespeare and early modern drama, The Shakespeare Institute, University of Birmingham

"A fascinating and hugely entertaining guide to Shakespearean mathematics."
—Prof. Sarah Hart, Gresham College professor of geometry, author of *Once Upon a Prime*

"Eastaway's playful and engaging book is packed with a dazzling array of historical facts, and is bound to excite the appetite of all Shakespeare junkies."
—Patrick Spottiswoode, founder, Shakespeare's Globe Education

"Instead of cleaving math and English in twain, Eastaway brings them together to surprise and delight the reader—and our understanding of Shakespeare's life and works is all the richer for it. The perfect book for students and teachers of math and English alike."
—Dr. Rebecca Fisher, The English Association

Much Ado About Numbers

Shakespeare's Mathematical Life and Times

ROB EASTAWAY

THE EXPERIMENT

NEW YORK

Much Ado About Numbers: *Shakespeare's Mathematical Life and Times*
Copyright © 2024 by Rob Eastaway
Page 205 is a continuation of this copyright page.

Originally published in the UK by Allen & Unwin, an imprint of Atlantic Books
Ltd. First published in North America by The Experiment, LLC.

The Experiment, LLC
220 East 23rd Street, Suite 600
New York, NY 10010-4658
theexperimentpublishing.com

The Experiment's books are available at special discounts when purchased in bulk
for premiums and sales promotions as well as for fundraising or educational use.
For details, contact us at info@theexperimentpublishing.com.

Library of Congress Cataloging-in-Publication Data available upon request

ISBN 979-8-89303-030-3
Ebook ISBN 979-8-89303-031-0

Cover design by Keenan
Author photograph courtesy of Rob Eastaway

Manufactured in the United States of America

First printing September 2024
10 9 8 7 6 5 4 3 2 1

CONTENTS

PROLOGUE

T he idea for this book was sown in Stratford-upon-Avon. Where else? The British Mathematical Association, an august body that has represented teachers and general mathematicians for over a hundred years, had decided to hold its 2022 annual conference in William Shakespeare's hometown. And that gave me the germ of an idea.

Stratford is of course still steeped in Shakespeare. Dominating the town is the huge Royal Shakespeare Theatre looking out towards the River Avon. The river is spanned by the fourteen-arch Clopton Bridge that Shakespeare would have traversed many times on his travels. Several half-timbered buildings survive from the sixteenth century, including the house in Henley Street where Shakespeare was born.

My friend Andrew Jeffrey and I had put ourselves forward to run a joint workshop at the teacher conference, as we have often done in the past. It would be our usual hour of light-hearted pick-and-mix mathematical curiosities.

But the location caught my imagination. Following the old rule that you're more creative if you put a constraint on yourself, I suggested that we theme our workshop on Shakespeare. Why not pick out Shakespearean quotes and link them to mathematical concepts? We could do a slot about the use of zero and call it *Much Ado About Nothing*, another one about fractions called *Henry the Fifth* (aka Henry the 20%), and dotted throughout there could be a *Comedy of Math Errors*. Cheesy, yes, but we knew our audience.

The curious thing, however, was that the more I hunted for superficial links between Shakespeare and mathematics, the more I realized that Shakespeare's world was filled with much deeper mathematical ideas, and that many of them are reflected in his plays.

If you'd asked me about math and Shakespeare before I started investigating it, I'd have dragged up hazy recollections about rhyming patterns in his verses, and also a mathematical rhythm in the way his lines were spoken, known as iambic pentameter. But it turns out that this is just the start of it. I have since uncovered numerous mathematical connections. As a result, a whimsical joke about Shakespearean math that was originally intended for teachers at a conference has ended up as this book.

I have been told by many people that in writing a book about Shakespeare I am stepping into a lion's den. There are thousands of academics, theatre producers, actors and historians across the world who have spent their lives studying Shakespeare, and there are countless books about him. His work has been analysed from just about every conceivable angle. Pick any topic – from horticulture to Harry Potter – and you'll find somebody who has investigated its connection with Shakespeare. I might, however, be the first to have looked at his life and work through the prism of math.

Before I embarked on this book, my knowledge of Shakespeare was typical of somebody who studied him at school (*Richard III* was my O Level play – I loved it) and who has then seen a couple of dozen productions on stage or screen, dotted through my adult life.

Happily, I now know considerably more about Shakespeare than I did when I started writing this book. It has been a joyful adventure. My research has taken me to the Royal Observatory in Greenwich, the real tennis court at Hampton Court, the graphite mines of

Cumbria, the inner sanctum of Oxford's Bodleian Library and the salvaged wreck of the *Mary Rose* ship in Portsmouth. I've spoken to historians of musical instruments, glove-making, taxation, art restoration, astronomy, stationery, card games and magic. And, of course, I have read plenty of books and spoken to plenty of Shakespeare experts too.

It is worth pointing out that, aside from the plays and poems that he wrote, we know remarkably little about Shakespeare the man, and what he studied or said. This lack of biographical detail means it is very easy to make wild speculations. I could, for example, claim that Shakespeare used to practise his times tables at breakfast, spend lunchtime totting up the accounts for the Globe theatre, and spend the evening winning money with loaded dice, in the confidence that nobody can be certain that I'm wrong. I will avoid such speculation, but, as you'll see, there is plenty of evidence that Shakespeare was aware of many of the mathematical ideas that were circulating in his lifetime, and that he incorporated many of them in his work.

You might ask 'Is this mainly a math book or a Shakespeare book?' I would reply that arguably it is a history book. Or maybe a science book. Or even a book about language, music and the arts. The truth is that it is a bit of all of them, because they are all connected. In the modern world, anyone who takes an interest in arithmetic or algebra might be described as a math person. But that is partly because we live in a binary world, where you are assumed to be interested in either the mathematical subjects or the arts, not both. In Elizabethan times, things were different.

The fact that leading mathematicians of that period, such as John Dee and Thomas Harriot, also experimented with optics, chemistry, map-making, seafaring and other diverse activities shows how blurred the edges were between different fields that are now distinct areas of study. It was a time when curiosity was rewarded because there were so many things still to be discovered and explained. This was the English Renaissance, and these were Renaissance men, living (for much of their lives) under a queen who was a Renaissance woman.

Shakespeare was born in 1564, at a time when there were huge advances happening in just about every sphere of life – in science, in the arts and of course in math itself. The radical astronomical ideas of Copernicus, who had suggested that the universe was centred around the sun rather than the earth, were a matter of huge religious and public debate in the late 1500s; arithmetic was becoming an essential skill for anyone involved in trade or business, be that manufacturing gloves (as Shakespeare's father did) or running a theatre; and music, regarded at the time as being a mathematical subject, was in the middle of its own revolution.

On the high seas, Walter Raleigh was one of the many adventurers opening up the world for exploration, which required increasingly advanced navigation skills. There was a huge growth in international trade as England developed a taste for exotic foods and spices, and at home the growth of mining and textiles was building the foundations for the Industrial Revolution that would finally explode a few generations later. Recreation was also taking off, with the invention of a plethora of sports and games, many of which provided a popular vehicle for gambling, which had a mathematics all of its own.

In this book I will be reflecting on the mathematics of Shakespeare's world as much as math in his plays: from games to astronomy, and from measurement to magic. In 1592, the playwright Robert Greene, who famously described William Shakespeare as an 'upstart crow', also called him a 'Johannes factotum'. That second description was particularly snide, as it is a variant of 'Jack of all trades'.

My sense of Shakespeare, however, is that he was an all-rounder in the best sense of the word, who was knowledgeable and curious about the world around him. And, while he wasn't necessarily a mathematician, he had a particular aptitude for numbers, and a mathematical mind.

A rounded education was assumed to include an in-depth understanding of the classics but also of mathematics. If you were interested in history, the sciences, languages or music, you were free to explore them, and to serendipitously find connections between them. That is the spirit in which I have written this book.

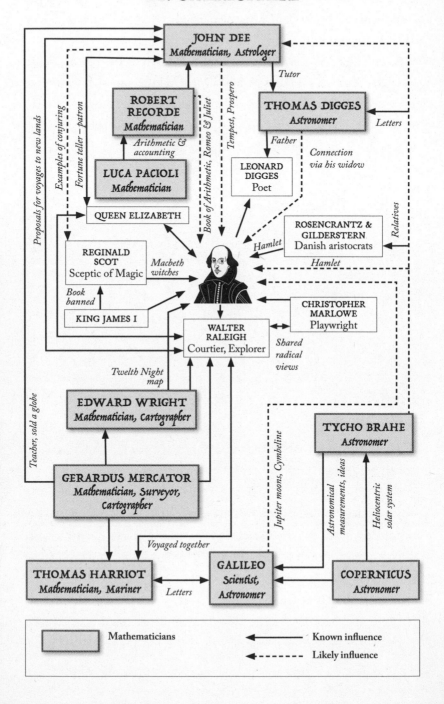

MATHEMATICAL INFLUENCES ON SHAKESPEARE

JOHN DEE
Mathematician, Astrologer

ROBERT RECORDE
Mathematician

Arithmetic & accounting

LUCA PACIOLI
Mathematician

QUEEN ELIZABETH

THOMAS DIGGES
Astronomer

Tutor

Father

LEONARD DIGGES
Poet

Letters

Connection via his widow

REGINALD SCOT
Sceptic of Magic

Macbeth witches

Book banned

KING JAMES I

ROSENCRANTZ & GILDERSTERN
Danish aristocrats

Relatives

Hamlet

Hamlet

Proposals for voyages to new lands

Examples of conjuring

Fortune teller – patron

Book of Arithmetic, Romeo & Juliet

Tempest, Prospero

WALTER RALEIGH
Courtier, Explorer

CHRISTOPHER MARLOWE
Playwright

Shared radical views

Twelth Night map

EDWARD WRIGHT
Mathematician, Cartographer

GERARDUS MERCATOR
Mathematician, Surveyor, Cartographer

Teacher, sold a globe

TYCHO BRAHE
Astronomer

Jupiter moons, Cymbeline

Astronomical measurements, ideas

Heliocentric solar system

Voyaged together

THOMAS HARRIOT
Mathematician, Mariner

Letters

GALILEO
Scientist, Astronomer

COPERNICUS
Astronomer

Mathematicians

Known influence

Likely influence

CHAPTER I

SHAKESPEAREAN NUMBERS

A PLAYFULNESS WITH NUMBERS

They doubly redoubled strokes upon the foe.

MACBETH

T he notion that Shakespeare's work might have any links at all with math might come as a surprise. In an era of Spanish Armadas, the arrival of tobacco, and the suppression of Catholics, you might think that math would have been the last thing on people's minds.

Yet what is striking is that Shakespeare's plays and poems are full of mathematical ideas, and in particular numbers. He quantifies just about everything: hours, years, distances, the size of an army: you name it, if Shakespeare can put a number to it, he generally does.

The more I've investigated the numbers he uses, the more impressed I've been by just how creative he is in expressing them. And, as you might expect, he is often poetic with numbers, too. One of my favourite examples appears in *Othello*, a tragic play about love and jealousy. In Act 3 Scene 4, Bianca is sorry that her lover Cassio has been away for a week. She begins:

What, keep a week away? Seven days and nights?

OK, we get it. But she decides to ram it home with another way of expressing just how long a week is:

Eight score eight hours?

Bianca is demonstrating some nifty mental arithmetic. She's letting the audience know that there are $7 \times 24 = 168$ hours in a week (or, as she puts it, 8×20 (a score), plus 8, which comes to the same thing).

Of course it's not Bianca who does this calculation; it's Shakespeare who has figured it out and then put the words into Bianca's mouth.

Did you already know that there are 168 hours in a week? I certainly didn't, until I worked it out. This would have been just as obscure a number fact in Shakespeare's time as it is today. It takes some mental agility to both calculate the number of hours in a week and then express it in a poetic turn of phrase. 'Eight score eight' sounds so much more elegant than 'one-hundred and sixty-eight'.

'Score' (meaning twenty) was one of Shakespeare's favourite numbers. The word comes from the Old Norse word *skor* meaning a mark or notch. One theory for why it came to represent twenty is that it was used for counting large numbers of, for example, sheep and making a notch in a stick for every twenty. Its meaning slowly changed to cover a reckoning or total amount, and by 1670 it had become a mark made to record a point in a game, which continues to this day.

The words 'sixty' and 'eighty' were very much in use in Shakespeare's time, but he hardly uses them, opting for 'three score' and 'four score' instead (sometimes hyphenated, sometimes a single word).

Keeping score

Use of the word 'score' (meaning 20) in Shakespeare.

	No. of appearances
Half a score (10)	1
One score (20)	6
Two score (40)	None
Three score (60)	6
Three score and ten (70)	2
Four score (80)	13*
Five score (100)	2
Six score (120)	1
Eight score (160)	2
Nine score (180)	2
Twelve score (240)	3

*'Four-score' also appears in *The Winter's Tale*, as 'Wednesday the four-score of April'. In this case, rather confusingly, 'four-score' is being used to mean twenty-four (4 + 20), not eighty (4 × 20). It's the only time Shakespeare uses 'four-score' to mean 24, and in other contexts it might have caused confusion. However, since April has only 30 days, the context here makes it clear that it must mean 24, not 80.

This use of multiples of twenty to express numbers was particularly popular in Elizabethan times. It's an idea that still lives on in the French numbering system. You may know that once French counting goes past 69 it becomes eccentric. The French 70 is 'sixty-ten' (*soixante-dix*) and 80 is 'four twenties' (*quatre-vingts*). Four score is the English equivalent of *quatre-vingts*.

Curiously, the French don't seem historically to have referred to 70 as 'three-twenties-ten' – but the English did. Shakespeare himself does so in *Macbeth* when the nameless 'Old Man' declares that he can remember 'three score ten' years of his life.

Shakespeare would have been well versed in the Bible and was no doubt familiar with Psalm 90, which includes the famous line that dictates the natural lifespan of an adult, the so-called allotted span. The King James Bible, published in 1611 (late in Shakespeare's career), puts it this way:

> The days of our years are threescore years and ten; and if by
> reason of strength they be fourscore years, yet is their strength
> labour and sorrow; for it is soon cut off, and we fly away.[1]

In other words, you can expect to live to 70, and if you somehow make it to 80, you'll be shuffling off your mortal coil pretty soon afterwards. Thankfully in modern times we've pushed a bit further beyond these limits.

There's plenty more number-play to be found in Shakespeare if you search for it. This includes, for example, referring to ten as 'half a score' (*The Taming of the Shrew*), and 'twice five' (in three plays). He also expresses fifty as 'half a hundred' (*Coriolanus*).

Shakespeare was also fond of compounding numbers and leaving the audience to imagine how big the result would be. In *The Tempest*, instead of describing himself as a complete idiot, Caliban says he is a 'thrice-double ass'. Thrice-double is six. Shakespeare does something similar in *The Merchant of Venice*, but on a bigger scale, when Portia suggests that a payment should be 'double six thousand and then treble that', which ends up as 36,000.

---

1   The Greek and Hebrew Bible texts referred to 'seventy' and not 'three score and ten', as did the first English (Wycliffe) translation of 1382, which says: 'The days of our years be seventy years'. However, by 1535, the Coverdale Bible was referring to 'three score and ten', as was the Geneva Bible of 1599. So, whichever English Bible Shakespeare read, the allotted lifespan would have been 'three score years and ten'.

And there's also what I would call exponential wordplay, which exploits the fact that if you keep doubling a number you reach a vast total very quickly. In *Macbeth*, a wounded captain is reporting how he saw Macbeth and Banquo fighting on the battlefield. He tells how

they doubly redoubled strokes upon the foe.

That sounds like a lot, but just how much is it? Double one cannon ball and you have two. Redouble and you have four. If you *doubly* redouble, you redouble twice, and four doublings means multiplying by sixteen. Shakespeare clearly liked this phrase because he'd used it before in the play *Richard II*, when John of Gaunt encourages Henry Bolingbroke to 'let thy blows doubly redoubled fall . . . on thy enemy'.

Ne'er the twain shall meet

Shakespeare mentioned the number *two* nearly 700 times, but he sometimes looked for creative alternatives, just as modern football reporters do. In the same way a journalist will say that a striker scored a 'brace' of goals, Shakespeare talks about a brace of greyhounds, or courtesans, or harlots. He also refers to a 'couple of pigeons'. But he had another word in his armoury that has disappeared from modern journalism: 'twain', an old English word for two. He uses it 47 times, including the line 'O Hamlet, you have cleft my heart in twain'. The word largely died out over the next 300 years, though in nineteenth-century Mississippi, boatmen measuring the river depth would shout a warning if the river was shallower than two fathoms, with the cry 'mark twain'. When the American writer Samuel Clemens was looking for a pseudonym, he decided to borrow this boating call.

And Shakespeare wasn't averse to a little numerical sleight of hand. Here is the Fool addressing King Lear:

> *. . . Leave thy drink and thy whore,*
> *And keep in-a-door,*
> *And thou shalt have more*
> *Than two tens to a score.*

In other words, live wisely and carefully, and you'll end up in profit, and two tens will end up being worth more than twenty. That's creative accounting for you.

HUGE NUMBERS

Ay, sir. To be honest, as this world goes, is to be one man pick'd out of ten thousand.

HAMLET

As well as being playful with numbers, Shakespeare also used them for dramatic effect. The more I've explored his plays, the more I've come to appreciate how he clearly loved the theatrical power of big numbers. His plays are full of them. The word 'thousand' appears over *three hundred* times in his work, most famously in his mention of the 'thousand natural shocks that flesh is heir to' in Hamlet's 'To be or not to be' speech. But that's small fry compared to some of the numbers he uses.

Before we go any further, it's worth considering what would have been viewed as a 'huge number' in Shakespeare's time. These days we get first-hand experience of big numbers all the time. We are used to travelling thousands of miles on holidays, we see crowds in the tens or even hundreds of thousands at football matches and rock concerts, and we routinely hear about houses valued in the millions. On top of that the government routinely spends billions and astronomical distances are often in the trillions (it's 23 trillion miles to Alpha Centauri). But the numbers in Shakespeare's world were much smaller. Few people would have had any experience of a distance of more than a hundred miles, and for most people an income of more than £100 per year would have been unheard-of riches. The population of London in 1600 was around 200,000 – roughly the capacity of two Wembley stadiums – but of course they'd never have all assembled in the same place.

The most tangible sense of a huge number would have been witnessing crowds of people. The biggest crowd gatherings would probably have been for the gruesome spectacle of public hangings. In London most public hangings took place at Tyburn, close to where Marble Arch is today. There were no officials tasked with counting the number in the crowd, but there are credible claims that several thousand would have turned up to watch the hanging of a notorious convict such as Sir Brian O'Rourke of Ireland. On 3 November 1591 (a Wednesday) he was found guilty of treason and towed through the streets to Tyburn. There, in front of a large crowd, he suffered the routine punishment of the age. First he was hanged and then (while still alive) cut open to have his bowels removed and thrown on a fire, before being beheaded and then having his body cut into quarters. Shakespeare was familiar with this ghoulish form of execution, because he refers to it in *King John* when Lewis, the French prince, is described as being so intensely in love that it has 'hanged drawn and quartered' him. Perhaps the 27-year-old Shakespeare was even in the Tyburn crowd that November day to see O'Rourke's execution.

The so-called Tyburn Tree, first erected in 1571, was used to carry out multiple executions. Hangings of notorious people are believed to have attracted crowds of several thousand.

But would a Tyburn spectator have been able to see how big the crowd was? I took the tube to Marble Arch to find the spot where the Tyburn Tree used to stand. The ground around there is actually quite flat, so you'd have needed to be on an elevated platform to get a real sense of what a crowd of thousands looked like. And the only people on an elevated platform were the hangman and the person he was about to execute, who probably had other things on their mind.

In fact if you wanted to witness thousands of people up close, the place to be was none other than Shakespeare's Globe theatre itself. It's been estimated that the theatre could hold about 3000 people, roughly double the capacity of its modern replica. Health and safety were not a big concern in Tudor times, and little thought was given to putting comfortable legroom between seats or aisles between rows. A Globe audience must have felt like a vast intimidating mob. *That* was what three thousand looked like.

What this all means is that anything above a few thousand was a massive, almost unthinkable number to your average Tudor, and so when Shakespeare referred to numbers this big, they'd have had quite an impact on his audience.

In Shakespeare's plays, the biggest numbers tend to crop up in three places: time, armies and money.

For example, it was common knowledge in Christian Europe that the world had been around for a very large number of years. According to Rosalind in *As You Like It*,

The poor world is almost six thousand years old.[2]

Perhaps Shakespeare believed this too. If so, it is probably the only number that he ever under-states (unless you are a Young Earth creationist).

---

2 The age of the earth had been calculated by going back through the chronology of the Bible. A figure of roughly 6000 years was widely accepted in Christian and Jewish cultures.

Meanwhile there are plenty of references to wealth and debts running into the thousands: 'three thousand crowns', 'ten thousand ducats' and so on (we'll find out more about the math of money in Chapter 4).

The biggest 'real' number of all to be found in Shakespeare is the fortune of one hundred thousand crowns that Ferdinand allegedly owes to the King of France in *Love's Labour's Lost*. That same fortune of one hundred thousand crowns is also to be found in *Richard II*. It's the equivalent of several million pounds today.

But Shakespeare's big numbers weren't limited to tallies of actual things. Numbers came into their own when they could be exaggerated to make a point. Occasionally Shakespeare left big numbers to the imagination, for example in the play *Pericles* the character Boult says he's described a woman down to 'the number of her hairs'. That's clearly a big number (the typical full head of hair has tens of thousands of follicles), but nobody's counting here.

Usually, however, Shakespeare does like to put a figure on it. When Henry VI is mourning the death of his uncle, Duke Humphrey, he wants to show the depth of his grief:

> *Fain would I go to chafe his paly lips*
> *With twenty thousand kisses, and to drain*
> *Upon his face an ocean of salt tears.*

Can we beat 20,000? Yes! Hamlet claims that his love for Ophelia is greater than that of 'forty thousand brothers'.

In *A Midsummer Night's Dream*, Oberon claims that Cupid's arrow should pierce 'one hundred thousand hearts', and in other plays we encounter one hundred thousand welcomes (*Coriolanus*), deaths (*Henry IV*) and flaws (*King Lear*).

Yet even one hundred thousand is not enough hyperbole for Shakespeare. He goes bigger still: Cranmer offers Henry VIII 'a thousand thousand' blessings (that's a million); Ferdinand offers Miranda a thousand thousand goodbyes; and Othello reckons that Bianca's singing makes her a thousand thousand times more contemptible than she would otherwise be, which means she must be very, very awful.

There are more than ten references to 'million' too.

But the biggest number of all to be found in Shakespeare's work is a little more disguised. It features in *Romeo and Juliet*, the play where melodramatic love and grief are at their most extreme.

The friar tells Romeo to escape to Mantua until his marriage can be announced in public. He reassures Romeo that, when he returns, Romeo will be greeted with vastly more joy than he'll feel when he leaves. But by how much more joy will that be, exactly? The answer is:

Twenty hundred thousand times more joy.

That's an increase in the level of joy by a factor of two million, which to a Tudor audience was an astronomically large number.

MUCH ADO ABOUT NOTHING

Nothing can come of nothing.

KING LEAR

W e've seen how Shakespeare enjoyed using huge numbers for dramatic effect. But there was another aspect of numbers that seems to have intrigued him even more.

It turns out that William Shakespeare lived through what was arguably one of the most important revolutions in mathematics of all time, in terms of its impact on the economy, education and society as a whole. Not only did Shakespeare know about this revolution, but he weaved its fundamental idea into his plays. It was a revolution that would make possible huge and rapid advances in knowledge across all of the sciences. And it would ultimately lead to the data-driven society that we live in today.

What was this revolution?

It was nothing. Or, to put it another way, it was the introduction of the digit '0' which, along with the digits 1 to 9, had made its way from India to the Middle East and then to Europe via the trading ports of Italy (including the merchants of Venice) and had arrived in England just a couple of decades before Shakespeare was born.

These 'Indo-Arabic' numerals (shortened to 'Arabic' from here on) were taking over from Roman numerals I, II, III ... and in the process were revolutionizing the way that numbers could be presented and manipulated. They had already had a huge impact on the way that business was conducted across continental Europe, because they enabled calculations to be done far more quickly, making trading much more efficient. Now England was joining the party.

Today we call the symbol for nothing 'zero', but that word wasn't in use in Shakespeare's time. *Zéro* first appeared in France around the year 1600 but didn't fully establish itself in the English vocabulary for another hundred years.

Instead, the word that Shakespeare's generation used for the zero digit was *cipher*, which (like zero) derives from the Arabic word *sifr* meaning 'nothing' or 'emptiness'. Today, a cipher is a code, a mystery that has to be solved, but that meaning evolved from the digit zero which to many people resembled something of a mysterious new code that had to be deciphered.

9	Nine.
8	Eight.
7	Seuen.
6	Six.
5	Fiue.
4	Foure.
3	Three
2	Two.
1	One.
0	Cipher

The 'new' digits from 9 down to 0 were introduced to the British public in Robert Recorde's 1542 book The Ground of Arts. *This image comes from the 1618 edition, and shows that, even two years after Shakespeare's death, the French word 'zero' was not in popular use in England.*

Cumbersome Calculations

You might be wondering why zeroes had such a profound impact on the way we handle numbers. Until the sixteenth century, numbers in England were written using Roman numerals. Seven letters, I, V, X, L, C, D and M, were used to express numbers up to the thousands. The year 1066, for example, would have been written as MLXVI. But Roman numbers could be cumbersome, especially when it came to doing arithmetic.

When Shakespeare's father John was a child in the late 1530s, if he wrote numbers at all it's likely that he would have exclusively used Roman numerals – his was the last generation to do so. If he wanted to use a written method to add numbers such as 12 + 25 he might have scrawled XII and XXV, and the letters could just be lumped together to get the answer XXXVII. (In practice, when he was doing calculations, he is more likely to have used his fingers, or counters on a table, and he'd then simply write the answer in numerals.)

Division and multiplication using Roman numerals could be harder. What's half of XLVI (46)? Without the aid of modern numerals and mathematical symbols, it's quite messy to work out. (The answer is XXIII.)

Multiplying would be equally cumbersome: twice 39 is 78, but with Roman numerals XXXIX doubles to LXXVIII. Expressing large numbers could require an ungainly number of characters. If you wanted to write 3874 you'd need twelve characters: MMMDCCCLXXIV.

This is why the Arabic decimal system popularized in England by Robert Recorde was so powerful. It introduced the idea of place value, so that a single symbol such as '5' could do much more work: it could be worth 5, 50, 500 or higher depending on which column it was placed in, with ciphers being used to indicate its value.

Perhaps the most famous example of Shakespeare using the word 'cipher' is in the Prologue for *Henry V*, when the Chorus (effectively a narrator), enters to announce that this small troupe of actors is about to recreate the glorious history of the battle of Agincourt. Just six or seven men will try to represent a huge army. He asks the audience to use its imagination: think of each actor as a zero. This little '1' (a 'crooked figure'[3]) can be turned into something vast, a million, when followed by six zeroes. The Chorus puts it this way:

> O pardon, since a crooked figure may
> attest in little place a million,
> And let us, ciphers to this great account
> On your imaginary forces work.

Shakespeare makes references to nothing-ness throughout his work. The word 'cipher' itself (in its zero sense) appears only five times, but he uses 'nought' or 'naught' (he seems to have been fairly flexible on spelling) 84 times, and the word 'nothing' itself crops up a staggering 590 times, not least in the title of one of his most popular comedies, *Much Ado About Nothing*.

The word 'nothing' in that play's title is actually a four-way pun. Its most obvious meaning is its 'numerical' one, that this play is about making a big issue out of something trivial. But, in Shakespeare's time, 'nothing' was pronounced in a way that made it sound like 'noting', which was a word for gossip (a feature of the play). Noting also referred to music, in the sense of making notes, and this musical meaning of noting is used in an exchange between two characters, Balthasar and Don Pedro:

---

3   Some scholars argue that the crooked figure refers to the '0' (zero), but I agree with those who think it refers to the digit 1, with the crooked figure a pun referring both to the digit and to the actor who is speaking.

Balthasar: *Note this before my notes, there's not a note of mine that's worth the noting.*

Don Pedro: *Why these are very crotchets that he speaks ~ Note notes, forsooth, and nothing.*

And finally, 'nothing' was also Elizabethan slang for a woman's private parts, because a woman had 'no thing' between her legs. So the title of the play *Much Ado About Nothing* is about trivialities (that tend to zero), gossip, music and sexual innuendo. Four meanings for the price of one word: that's Shakespearean punning at its best.

Shakespeare even refers to the *shape* of the zero digit. In the Chorus prologue for *Henry V*, when describing themselves as ciphers, the characters also refer to how they are performing in 'this wooden O', 'O' being both a letter and a digit. In this case, the wooden O was referring to the Globe theatre, which was (and is) an O-shaped theatre.[4]

This abundance of references to 'zero' suggests that Shakespeare was very much taken with this novel way of expressing numbers. And why wouldn't he be? Any creative and curious person would have been intrigued by the way something so small and simple could radically change the way people thought, worked and communicated. It was in its way the Tudor equivalent of the mobile phone.

Ciphers had an almost paradoxical role as both adding huge value, yet being worth nothing at all unless they were preceded by one of the other digits – an idea that clearly appealed to Shakespeare. For example, when making fun of King Lear after he has given up his kingdom, the Fool declares that Lear is now an 'O without a figure', in other words a zero without a figure in front of it, which means he is nothing:

---

4   Henry V was first performed in 1599 at the Curtain theatre in Shoreditch, which was rectangular, a few months before the Globe opened in Southwark. It's believed that Shakespeare almost certainly changed the prologue to add in the wooden O reference after the play moved south of the river. The reconstruction that was opened in 1997 recreates what the original theatre is believed to have looked like.

Thou wast a pretty fellow when thou hadst no need to care for her frowning; now thou art an O without a figure. I am better than thou art now; I am a fool, thou art nothing.

And in *The Winter's Tale*, Polixenes wants to express to his old friend Leontes just how immensely grateful he is for the hospitality he has received over the last nine months. He's just one man, but he can multiply his thanks by adding more zeroes.

And yet we should, for perpetuity,
Go hence in debt: and therefore, like a cipher,
Yet standing in rich place, I multiply
With one 'We thank you' many thousands more
That go before it.

Shakespeare also neatly exploits the mathematical fact that if you divide the number 1 by a very large number, the answer is very close to zero. In *Romeo and Juliet*, the handsome Paris reveals that he's interested in marrying Juliet. Her father Lord Capulet isn't happy at the idea, so he throws a party to which many attractive young women are invited. That way, Juliet becomes just one of many, and (if you'll excuse some modern math terminology) the probability p that Paris still finds her the most attractive is now 1 divided by the number of female guests, N, or $p = 1/N$, where N is large, and hence p tends towards zero.

Or, as Shakespeare puts it, rather more eloquently:

Which on more view, of many mine being one
May stand in number, though in reckoning none.

All of which goes to show that Shakespeare made much ado about nothing.

CHAPTER II
SCHOOL LIFE

MATH AT GRAMMAR SCHOOL

Then the whining schoolboy, with his satchel
And shining morning face, creeping like snail
Unwillingly to school.

AS YOU LIKE IT

Where did Shakespeare learn his math? Like all of us, he would have learned some math at school. Between the ages of five and seven he might have attended what was known as a petty school, run by a teacher or housewife at their home, where in addition to learning the basics of reading and writing, he would have learned to count and to write numbers. From there he would have gone to grammar school where his formal education would begin.

The fact that Shakespeare went to school at all made him special. Only a tiny fraction of children, fewer than one in a hundred, were formally educated in his day. Boys who didn't go to grammar school at age seven would be starting an apprenticeship in their father's trade or working in the fields. Girls would generally be set to work around the home.

The reasons why Shakespeare got the huge advantage of a formal education were twofold:

1. His father was an affluent glovemaker and a member of the town council (an alderman), making him an important member of the local community.
2. He was a boy. (We'll see what happened with girls later.)

There's no written record of where Shakespeare went to school, but he surely attended the grammar school in his home town of Stratford-upon-Avon, a three-minute walk from his house. In fact, you can still visit the classroom where Shakespeare spent a large part of his youth. I did just that. It's an evocative experience to sit in the very room where he learned and recited poetry, and most likely performed Greek dramas. The Elizabethan school was a single upstairs classroom in a long, half-timbered building. There was no blackboard, just a throne-like chair where the teacher would sit. Almost all the teaching would have been oral, without the visual aids that these days we would regard as essential.

Shakespeare's classroom. Boys sat on benches facing the centre, with the master on the high chair at the front of the class.

In Shakespeare's time there would have been about 40 pupils, seated on benches (known as *forms*) facing in to the centre of the room rather than facing the front. All age groups were in the same room, seven year olds would be on the first form, eight year olds the second form,[5] and so on. A single master and an assistant (known as an usher) were in charge. Older boys were expected to help with the teaching, by passing on their learning to the younger ones. It sounds very worthy and collaborative in principle, but in practice? I dread to think.

There might have been a little math in Shakespeare's curriculum, but it would have been only a very small part of his instruction.

It might come as a surprise that somebody attending a grammar school didn't spend hours a week studying math. What was he doing for all that time in the classroom? The answer, in short, is that he was studying Latin. Almost constantly. From 7am to 5pm every day, including Saturday, except for a two-hour lunch break (in which he would still be expected to speak in Latin when playing, or face corporal punishment).

The breakdown of a typical school timetable for a twelve-year-old in an Elizabethan grammar school is shown in the table.

	MONDAY	TUESDAY	WEDNESDAY	THURSDAY	FRIDAY	SATURDAY
7am -11am	Latin Text	Latin Poetry	Latin Text	Latin Text	Repetition of the week's learning	Latin exam
11am -1pm	B	R	E	A	K	
1pm-5pm	Latin grammar	Latin grammar	Latin grammar	HOLIDAY	More repetition	Latin exam Arithmetic

---

5   This is still used to describe a year group, most commonly the sixth form (aged 16–18).

In practice those hours would have varied across the year. In winter the timings were dawn till dusk – though boys were expected to provide their own candles if it got dark. Elizabethan pupils spent roughly twice as many hours in lessons over a year as their modern counterpart.

Only boys with an aptitude for learning would be allowed to attend school. Even so, if you try to picture a modern-day twelve-year-old having to deal with even one day of that curriculum, you can begin to imagine what drudgery it must have been for many children. Even the number-phobes would probably have been thrilled to have some multiplication exercises, just for a bit of variety.

Why didn't schoolboys riot? In part, they didn't know any different; this is just what school was. But what kept them in their place was the constant threat of being beaten. Shakespeare's fellow playwright Ben Jonson said that one of his schoolmasters had spent his days 'sweeping his living from the posteriors of little children'. No wonder the schoolboy crept 'like snail unwillingly to school'. The school regime hardly bears thinking about.

The math that Shakespeare did encounter was arithmetic. This would have been taught ad hoc, perhaps on Saturday afternoons, and possibly by a visiting tutor. It would have been little more than adding, subtracting and multiplying. No algebra or geometry was taught at school.

What about Pythagoras, that staple of modern secondary school math whose theorem enables us to calculate one side of a right-angled triangle from the other two? Shakespeare does mention Pythagoras four times in his plays, but this is nothing to do with the famous theorem. Instead, it reflects what Shakespeare would have learned from studying Ovid's *Metamorphoses* in his Latin lessons. Pythagoras believed that all animals had immortal souls that could be reincarnated as humans.

Shakespeare was clearly taken with this idea, because he mentions it in three plays, including *Twelfth Night*:

Fool: What is the opinion of Pythagoras concerning wildfowl?

Malvolio: That the soul of our grandam might haply inhabit a bird.

Fool: What thinkest thou of his opinion?

Malvolio: I think nobly of the soul, and no way approve his opinion.

Alas on the square upon the hypotenuse Shakespeare remains silent.

THE BOOK OF ARITHMETIC

*A plague o' both your houses! Zounds, a dog, a rat . . .
a rogue, a villain, that fights by the book of arithmetic!*

ROMEO AND JULIET

The mention of a 'book of arithmetic' in *Romeo and Juliet* shows that
Shakespeare was aware of such a book, though it is unlikely that he
had his own when he was at school. Books were far too expensive for
there to be textbooks for every child, but whoever taught arithmetic
to Shakespeare almost certainly used *The Ground of Arts* by Robert
Recorde as his guide – it's the book that introduced ciphers, which we
met in Chapter 1. First published in 1542, it was one of the earliest
arithmetic books in English, and became the pre-eminent textbook for
a hundred years.

Robert Recorde was born in Tenby in South Wales around 1510,
where his father was a merchant. Robert went to Oxford University
where he specialized in medicine. From there he went to Cambridge,
where he probably taught and practised as a doctor for a while. However,
he was also passionate about mathematics. Recorde was well aware of
the new number system and arithmetical methods that had taken hold
in the rest of Europe, and he thought it was vital that England should
catch up. In his words:

> Sore oft times have I lamented the unfortunate condition of
> England, seeing so many great clerks arise in sundry parts of the
> world, and so few to appear in this our nation.

This was despite the fact that, in his opinion, 'Englishmen are
inferior to no men.' He put the country's innumeracy down to the

Englishman's contempt for learning, and in particular (it seems), a contempt for learning math. You might hear much the same comment today.

There's a hint of this attitude in *Love's Labour's Lost*, when Don Adriano de Armado seems to boast about the fact that being able to do arithmetic is beneath the likes of him; it's better suited to inn-keepers who have to tot up figures for a living. When he is asked to calculate 1 times 3, he replies:

I am ill at reckoning. It fitteth the spirit of a tapster.

Recorde set about writing a book that would be accessible to the layman, laying out in detail the methods that we still use today for adding up in columns, subtraction by 'borrowing', long division using what we now call the bus stop method, and much more besides. To give the book popular appeal, he added humour, often in the form of rhymes. Early in his book, he explains why being innumerate is a huge handicap in life. As he puts it:

If number be lacking
It maketh men dumb
So that to most questions
They must answer mum.[6]

---

6   'Mum' here means silent or unable to speak, as in the expression 'mum's the word', which
    was first used by Shakespeare in *Henry VI Part 2* (his version was 'no words but mum').

Farmers were among those who would have needed help with handling figures. In *The Winter's Tale*, Shakespeare pokes fun at the innumeracy of the common man. The Clown character (a shepherd's son) arrives on stage looking baffled.

Let me see: every eleven sheep yields a pound and odd shilling; fifteen hundred have been shorn. What comes the wool to? [7]

The calculation is too much for him:

I *cannot do it without counters*

he says, in frustration – counters being the abacus system that was in common use at the time.

There was a vast audience that could benefit from Robert Recorde's tips. As well as introducing people to Arabic numerals and the idea of decimal 'place value' with columns for units, tens, hundreds and so on, Recorde also dedicated whole chapters to the new arts of addition, subtraction, multiplication and division.

This included a written method for doing multiplication by numbers between 6 and 9, in other words a trick for doing the 6, 7, 8 and 9 times tables, which was similar to a finger method being used at that time.

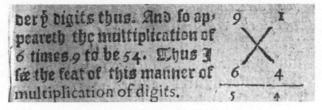

Recorde's multiplication method.

If Shakespeare wanted to learn how to multiply 8 × 6, for example, he might have been taught to do it as follows. Start by putting the two numbers in a column on left, with diagonal lines to the right. Now, on the right, put the numbers that add to these to make ten. (Primary school children today would call these 'number bonds to 10'.) As 8 + 2 = 10, we put 2 top right; 6 + 4 = 10 so we put 4 bottom right:

Now for the sleight of hand. To get the units digit of the answer, multiply the two numbers in the right-hand column: 2 × 4 = 8. And to get the tens digit of the answer, choose a diagonal (it doesn't matter which) and find the difference between the numbers on the diagonal: 8 − 4 = 4 (and so does 6 − 2).

So the answer to 8 × 6 is four-eight or 48. You might like to test it out with some other numbers, such as 7 × 9. You should get the answer 63.

This technique was only intended to be used for numbers between 6 and 9 (Recorde reckoned the smaller times tables were easy and could simply be worked out with fingers).[8]

---

8   If you want to push the boundaries of this criss-cross technique, you can experiment with what happens when you try 6 × 5, or 12 × 13. It still works, but you have to know about carrying numbers across columns and, in the case of 12 × 13, you need to be happy with multiplying negative numbers – a concept that was unfamiliar in Shakespeare's time.

This might seem a lot of work for individual times table calculations, but with practice it comes quickly, and the idea was that once you'd done it enough times the answer would stick in your memory.

Whichever way Shakespeare learned his times tables, it seems that he had indeed memorized them. In *Coriolanus* he describes eighteen as 'thrice six'. Whether his audience knew the answer is another matter – few of them would have been formally taught multiplication, though perhaps they would have learned their 6 times table through everyday handling of dozens and half-dozens (of eggs, for example).

To illustrate his arithmetical techniques, Robert Recorde peppered his book with word problems, just as today's math teachers would do. As it happens, in a problem slightly reminiscent of the one that Shakespeare's shepherd boy struggled with, one of Recorde's examples is about sheep:

> There are 29 men and each man hath 264 Lambs. The question is, how many Lambs they have in all.

(Robert Recorde's math problems were just as ludicrously implausible as the ones encountered every day in modern math textbooks.)

To work out the number of lambs, Recorde set out the calculation 29 × 264 like this.[9]

---

9   The 29 in Recorde's calculation appears to be crossed out. Presumably after multiplying by a digit, Recorde liked to cross it out as a check that no digits had been missed out.

The working looks similar to the long multiplication method that you may have learned at school. Starting at the right-hand side with the units digits, 9 × 4 = 36, write down the 6 and carry the 3; 9 × 6 = 54, add the carried 3, giving 57, write down the 7 and carry 5; 9 × 2 = 18, add the carried 5, giving 23, making a total of 2376 for the first line of the multiplication. And so on.

However, notice that the 'carried' numbers haven't been written down. There's no sign of a little 3 (from 36) in the tens column, or little 5 (from 54) in the hundreds column. Presumably the solver in Shakespeare's day was expected to hold this value in their head. Nor is there a zero (cipher) after 528. Recorde didn't think this was necessary.

But there's something rather more important that's missing: the multiplication sign itself. At first glance how do you know this is multiplication and not addition? Presumably only through the context.

It would be almost ninety years after Recorde's book was published before an Englishman, William Oughtred, in 1630 'invented' the multiplication sign resembling an × that we use to this day, long after Shakespeare died. Where did Oughtred get the idea for that × symbol? He must have been aware of Recorde's times table technique with the giant × in the middle. A smaller version of it would be a natural way to represent the operation.

But that came later. When Shakespeare multiplied, he had no symbol to do it with.

THE SEVEN LIBERAL ARTS

Debate logic with your acquaintances, And practise rhetoric in your common talk; Use music and poetry to quicken your mind; Turn to mathematics and metaphysics when you can stomach it.

THE TAMING OF THE SHREW

Shakespeare's education ended abruptly at the age of thirteen or fourteen. His father's business was in trouble, and either there was no money left for school fees, or his father needed William at home to help bring in money. Either way it meant that Shakespeare could not go to university, which was the natural destination for somebody with his talent. As a result, he missed out on a substantial part of the math education that some of his peers, such as the playwright Christopher Marlowe, received.

The only two universities in England were Oxford and Cambridge, and typically students enrolled at the age of fourteen. It was very much a continuation of school, with strict discipline and very long hours. Teaching was done in Latin, and almost all of it was oral. As with school, there were no blackboards. The main difference was that boys would be living in lodgings rather than at home, and would therefore be at the mercy of negligent landlords and the rough and tumble – or outright violence – of daily life.

In Tudor times, education across Europe was based on the seven[10] so-called Liberal Arts. These seven areas of learning were split into three that you studied at school (the *Trivium*) and four that would be studied if you went to university (the *Quadrivium*).

---

10 This is the first of several appearances of the number 7 in this book. Seven had a huge influence on the way Tudor England viewed and managed their world, and that influence remains to this day.

The modern word 'trivial' comes from the fact that the Trivium were the basic school subjects, though they seem far from basic now. The three parts of the Trivium were grammar, logic and rhetoric, all wrapped up in a relentless study of Latin literature day after day for several years at school, as we have seen. 'Logic' was learning how to make a sound argument based on robust statements from which inferences could be made. It's a process familiar to any mathematician, though Shakespeare's version would have involved no mathematical symbols. All of this was learned by translating and memorizing work by Horace, Cicero, Livy, Ovid and the other great classical writers and poets.

The reason why Latin was deemed so important is that in Elizabethan England you needed it to be able to read almost any book and to read or write any legal document. The study of Latin classics also taught you skills in reasoning and in arguing a case – vital skills if you were to go on to the top professions of the time, in the law, the church or the military. And, as it turns out, these were pretty handy skills if you wanted to be a leading playwright too.

The 'non-trivial' stuff – the university level Quadrivium – was based around math. The four mathematical subjects were:

Arithmetic (a more advanced form of
    what had been learned at school)
Geometry
Astronomy
Music

Music might come as a surprise but, as we shall see later, music was regarded as being as much a part of mathematics as fractions and ratios are today.

Illustration from Margarita Philosophica *(literally 'Pearl of Wisdom'),
an early sixteenth-century university textbook by Gregor Reisch, a
German monk, depicting the 'new' arithmetic using Indo-Arabic numerals
competing with the old arithmetic on a counting board. The use of
Arabic numerals in continental Europe was decades ahead of England.*

While Shakespeare didn't get the opportunity to formally study
these topics, his numerous references to astronomy and music suggest
he knew quite a lot about both – much as many of us might have a
lay understanding despite not studying these subjects at school. There
were no public libraries, but Shakespeare would have been able to buy
books at shops in London, many of which were street stalls around
St Paul's Cathedral. A well-bound book might cost him a day's

earnings, so it would be a treasured possession that he would want to hold on to for years.

Not everyone approved of this new-fangled, highfalutin education. In *Henry VI Part 2*, a clerk called Emmanuel is brought in front of the rebel leader Jack Cade. When it's discovered that Emmanuel can 'write and read and do accounts', Cade sends him away to be executed. Later on, Cade has angry words for Lord Say, before having him executed too:

Thou hast most traitorously corrupted the youth of the realm in erecting a grammar school; and whereas, before, our forefathers had no other books but the score and the tally, thou has caused printing to be used.

GIRLS AND MATH

To instruct her fully in those sciences,
Whereof I know she is not ignorant.

THE TAMING OF THE SHREW

Almost no girls attended school beyond the age of seven in Tudor times, in England at least.

Girls did get some instruction beyond that age, but their education was generally in the practicalities of the home rather than the wider intellectual disciplines that boys were exposed to.

There were exceptions, most notably Queen Elizabeth I herself. Elizabeth had a very broad education and by all accounts she excelled – she was fluent in several languages, for example, and she probably had some tuition from John Dee, one of the leading mathematicians of the time, who would have instructed her in astronomy and some elements of mathematics too.

And it was not unheard of for private tutors to give lessons in mathematics to girls in affluent households. In *The Taming of the Shrew*, Hortensio wants an opportunity to court Bianca in private. Bianca is the daughter of a wealthy businessman, who is a gentleman but not a member of the nobility. Hortensio disguises himself as a tutor who is 'cunning in music and the mathematics' and arrives to instruct Bianca fully in 'those sciences whereof I know she is not ignorant'. In other words, Bianca has already had some math instruction.

One of the most impressive examples of a Shakespeare contemporary who had strong numeracy skills was another Elizabeth, the countess of Shrewsbury, better known as Bess of Hardwick. Bess grew up in a family of minor gentry in Derbyshire. She might have attended a petty school (or dame school) as a young girl and learned about numbers then, but after that her education would have been entirely gained 'on the job'. She is famous for having married four times, to men who were increasingly wealthy, and ending up as England's second wealthiest woman after the queen. However, her wealth was only partly down to inheritance. It's clear that behind the scenes she was keeping a close eye on the numbers. With her second husband William Cavendish she bought Chatsworth House, and as head of the household she began keeping track of the accounts of the estate. In fact it's been estimated that about 90% of the records in the accounts for Chatsworth were written in Bess's hand, the rest being written by her husband Sir William and her steward Francis Whitfield.

She also did the adding up. The kitchen accounts from 13 November 1552 onward were kept by a steward but were totted up and signed by Bess every day. She clearly had a sound grasp of arithmetic.

Meanwhile in the households of the wider population, it was often the woman of the household who looked after the bookkeeping, which meant that basic numeracy was essential, and mothers no doubt passed their knowledge on to their daughters. Shakespeare had two daughters, Susanna and Judith. It's believed that Susanna was educated and literate, as she could write her name and she was joint-executor of his will, along with her husband John Hall, a physician. The Halls inherited Shakespeare's property, and the management of rent and bills might well have fallen to Susanna.

There's no evidence that her younger sister Judith was as well educated, and she may not have had skills in doing any sort of written accounts.

However, there was always mental arithmetic, which didn't require any writing. Shakespeare suggests that figuring things out in your head was a last resort. In *Troilus and Cressida*, Thersites jokes about how anxious the commander Ajax is looking, walking around like a peacock and biting his lip

> *. . . like a hostess that hath no arithmetic but her brain to set down her reckoning.*

Although intended as a criticism, this comment suggests that it wasn't uncommon for a 'hostess' (a woman working in an inn) to have a decent ability in mental arithmetic.

CHAPTER III

SPORT AND GAMES

ANYONE FOR TENNIS? OR FOOTBALL?

Like a school broke up,
Each hurries toward his home and sporting-place.

HENRY IV PART 2

Those Elizabethan schooldays were long, but there was still some time for play, even for scholars. There was, after all, a two-hour break in the middle of the school day. It's often through games and recreation that people of all ages acquire mathematical skills and strategies, particularly in games of chance. Perhaps it is through playing as a child that Shakespeare picked up some of his mathematical ideas.

In an era where so many people lived hand to mouth, it seems remarkable that any adults had time to spend on recreation. Financial insecurity, plague and famine were always around the corner. And yet play really *was* the thing.

Not only were thousands of Londoners attending the theatre every day – in the middle of the afternoon – but on street corners, in homes, fields and particularly in pubs, people were playing games and, to a lesser extent, sports as well. No doubt Shakespeare was one of them.

Shakespeare mentions about fifty different games and sports in his work. Some of them are still familiar to us – billiards, blind man's buff, archery and chess, for example – but others are long forgotten games and pastimes, including a couple that involve rolling marbles: one called 'troll-my-dames' (in *The Winter's Tale*) and the other called 'cherry-pit' (*Twelfth Night*). As a reminder of what brutal times these were, another popular recreation was bear-baiting, which is alluded to in several plays including *Macbeth* and *Julius Caesar*.

Shakespeare also mentions football and tennis, though football in particular was very different from the game we know today. There's a clue to what the Elizabethan version of the sport was like in *King Lear*. Lear has a confrontation with Oswald which he likens to a game of football:

Lear: Do you bandy looks with me, you rascal! (Striking him).
Oswald: I'll not be strucken, my lord.
Lear: Nor tripp'd either, you base football player (Tripping up his heels).

In short, although the aim of football was nominally to get a ball into a goal, it was really little more than recreational violence which often led to broken bones. For most players, the ball was largely irrelevant. Shakespeare might have played some football in his lunch breaks (kicking around an inflated pig's bladder or a stone), and might even have participated as an adolescent in a free-for-all match in the streets of Stratford-upon-Avon. There were no formal rules, no league tables, no statistics; in fact, from the point of view of this book, nothing of mathematical interest whatsoever – unless you're interested in the tally of visitors to an Elizabethan Accident and Emergency unit. Except of course there was no A&E – in fact there was no 'hospital' at all (in Stratford, anyway). If you broke a leg, friends or family might rush to find the town's barber surgeon, whose expertise extended beyond cutting people's hair to cutting open their bodies. He would probably have had experience dealing with wounded soldiers, so had some skill in stitching up wounds and applying splints to broken limbs. If the patient was lucky, the surgeon might even have an anaesthetic known as dwale, a concoction whose ingredients included lettuce, vinegar, hemlock and (perhaps the key ingredient) opium.[11]

By coincidence, the *King Lear* football quote also alludes to tennis. 'Bandy' means to hit to-and-fro and was the word used to describe a rally in tennis. Tennis was a much more tactical and mathematical game, but it was played almost exclusively by the aristocracy. The courts were indoors, and expensive to build. Shakespeare might well have witnessed a match, but it's unlikely he ever played tennis. The version being played in Shakespeare's time survives, and is these days known as 'real tennis', not (as is often claimed) because 'real' is an old word for 'royal', but simply because when lawn tennis came along in the late nineteenth century, the original version was described as 'real' by the traditionalists who were not into this new outdoor grass nonsense. There are still a number of real tennis courts around England and Scotland, the most famous being in Hampton Court.

The rules of real tennis are extremely complex. As part of my research for this book, I treated myself to an hour's coaching from Scott Blaber, a professional at Hampton Court. Despite Scott's patient introduction to the rules, my grasp of what a player needs to do to actually win a point is still shaky. The scoring itself, however, is remarkably familiar. Players compete for points, games and sets. Points are scored as 15, 30, 40 and then game or, at 40–40, deuce and then advantage. First player to six games wins a set. First to two (or sometimes three) sets is the winner.

---

11 In *Othello*, Shakespeare hints at the anaesthetic properties of opium: 'Nor poppy ... nor all the drowsy syrups of the world, shall ever medicine thee to that sweet sleep that thou (had) yesterday.'

The sport and its scoring system almost certainly originated in France, in a game known as *jeu de paume* ('game of the palm'), which in its earliest form was a hand-ball game, with the hand being used to hit the ball. It evolved into a game that used rackets. It is often claimed that scoring was originally done on a clock face, with the hand moving around by a quarter after each point – to 15, 30, 45 (which later got rounded to 40) and then game. It's an appealing theory, but since tennis scoring almost certainly predates clock faces by a couple of hundred years, it's almost certainly wrong. As I mentioned earlier, counting systems based around 60 were popular in ancient times, and they lived on in France for longer than in most other countries. So perhaps tennis scoring evolved simply from dividing 60 into four equal parts, with no thought of minutes and time at all.

Shakespeare makes no mention of tennis points, but he does allude to some of the tactics. In *Henry V*, relations are frosty between the French dauphin and the young King Henry, who has his eye on reclaiming control of northern France. The dauphin sends his ambassadors with a gift for Henry – a chest of tennis balls. The underlying message is:

Men playing Jeu de Paume, 1608.

'hey young man, stop messing with us, here are some toys for you to play with'. Alas for the French, this teasing proves to be a big mistake.

Henry's riposte is a speech filled with sarcasm, seething anger and a thirst for revenge – all in the form of a tennis metaphor. It begins as follows (the tennis terminology is underlined):

> We are glad the Dauphin is so pleasant with us;
> His present and your pains we thank you for:
> When we have march'd our <u>rackets</u> to these <u>balls</u>,
> We will, in France, by God's grace, <u>play a set</u>
> Shall strike his father's crown into the <u>hazard</u>.
> Tell him he hath made a <u>match</u> with such a wrangler
> That all the <u>courts</u> of France will be disturb'd
> With <u>chaces</u>.

And so it goes on. In short, Henry is telling the French that the English are going to come and smash them around the park in battle, and sure enough at Agincourt that's exactly what they do (while crying 'God for Harry, England and St George!').

Rackets, balls, sets, courts and matches are terms that are still used in lawn tennis, but 'chace' and 'hazard' are not. A chace (or chase)[12] is the line along which the ball bounces for the second time, and is used as a target for scoring the next point (I told you it was complex). A hazard in real tennis is a small recess in the wall at the receiver's end of the court where the ball is not retrievable, so a winning point is scored.

This use of the word 'hazard' in tennis is interesting. In Shakespeare's time, hazard was beginning to take on its modern meaning of something that involves risk or danger. The hazard end of the tennis court is the end at which there's the greatest risk of losing a point. But the word 'hazard' was originally linked to a much more mathematical game.

---

12 Some people such as Hilary Mantel have claimed that this is the origin of the expression 'cut to the chase', though this is disputed.

HAZARD AND RISK

And by the hazard of the spotted die
Let die the spotted.

TIMON OF ATHENS

The word 'hazard' comes from the Arabic word *Al-zahr*, meaning dice. If you mentioned the word 'hazard' to the average Elizabethan in the street, they would probably think of the popular dice game that went by that name. The dice game Hazard swept across continental Europe in the fourteenth century (in Italy it was called Zara). Henry VII had tried unsuccessfully to ban it, as had his son Henry VIII. But, despite that, or maybe because of it, the game thrived. By the time Elizabeth I was on the throne, it had become the most popular game in Shakespeare's England.

And why did the authorities dislike it so much? Because it was a gambling game. And, to add to its evils, Hazard was widely played in pubs, which were a magnet for some of the most unsavoury characters in society – whores, thieves and con artists. To get an appreciation of what an alehouse was like in Shakespeare's time, and purely for research purposes of course, I paid a visit to the George Inn in Southwark, which dates back to the sixteenth century. With its low half-timbered ceilings, small rooms and dark oak tables, it doesn't take too much imagination to picture the room filled with tobacco smoke and fellows crowded round a table rolling dice. Shakespeare very likely visited this alehouse himself – in *Henry VI* he bases one of the characters, Jack Cade, in the White Hart Inn, which was next door to the George and Dragon (as it was then known), and surely Shakespeare would have been familiar with the neighbouring pubs too.

Like modern-day poker and roulette, the main pleasure associated with the game Hazard came from placing bets on the outcome.

This more common use of the word 'hazard' to mean gambling also appears in the play *Henry V*. As the French warm up for battle, the nobleman Lord Rambures challenges his men:

Who will go to hazard with me for twenty prisoners?

In other words, 'I bet I will capture twenty of the English. Who wants to take me on?'

The game of Hazard itself involved rolling either two or three dice (there were no fixed rules). In its simplest form, the players merely bet with each other as to what total would turn up with the next roll of the dice. Since dice were numbered 1 to 6, just as they are today, in the two-dice version, this would mean betting on a total in the range 2 to 12, while for three dice the total would be between 3 and 18.

To understand a dice game and play it optimally, you need to be a master of probability theory. In Shakespeare's day, however, this area of mathematics had not yet been developed. Things had at least moved on from the ancient Roman view. For Roman dice players, the outcome was already written in the stars, and there was therefore no point in trying to apply any skill.[13]

Experienced Hazard players who were gambling their savings every night would have been very aware that three dice would more often add to a total of ten than a total of, say, three, but quite why this was so, and how much more likely a ten would be, was still unknown. No doubt there were plenty of theories, but probability is notorious for its counterintuitive outcomes.

The modern understanding of probability did not become public knowledge until two French mathematicians by the names of Blaise Pascal and Pierre de Fermat published their theories late in the mid-seventeenth century. In private, however, there were people who

---

13 In *Antony and Cleopatra*, Antony says of Julius Caesar: 'The very dice obey him', suggesting that only the powerful and supernatural can influence the outcome of a dice game.

Sixes and sevens

It is believed that the game of Hazard was the origin of the phrase 'to be at sixes and sevens', which means to be in a state of confusion.

Michael Quinion, a British etymologist, reported on his website that it's thought the expression was originally: *to set on cinque and sice* (from the French numerals for five and six). Five and six were believed to be the riskiest numbers to shoot for ('to set on') and anyone who tried for them was considered foolish. The three-dice version of Hazard was common in continental Europe, so totals of five and six would certainly have been very unlikely (and risky) choices.

The phrase 'to set the world on six and seven' was used by Chaucer in the mid 1380s, and seems from its context to mean 'to risk one's life'. Shakespeare uses a similar phrase in *Richard II*, which is closer to the modern meaning of being in disarray:

But time will not permit: all is uneven,
And every thing is left at six and seven.

already knew the odds and how to exploit them long before Pascal and Fermat. In fact, we know that the secret of how to win at Hazard was uncovered as early as 1620, just a few years after Shakespeare's death. And who cracked it? It was the astronomer and mathematician Galileo Galilei who had noticed that friends of his patron, the Archduke of Tuscany, were struggling to understand the math behind rolling three dice. (We will meet Galileo again in Chapter 7.)

The particular problem that Galileo wanted to solve was this: When rolling three dice, there would appear to be as many ways of getting a total of 9 as there are of getting a total of 10. And yet a score of 10 seems to come up more often. Why?

Galileo's elegant explanation of why it is better to bet on 10 (or 11) than on 9 or 12 when rolling three dice is explained in the box at the end of this section.

Curiously, although Galileo wrote up his findings in 1620, they were not published until 1718, long after his death. Why the hundred-year wait? We don't know for sure, but it's easy to hazard a guess, if you'll excuse the pun. Knowing that the odds of getting 10 were higher than the odds for 9 was hot knowledge that would give a significant financial advantage to anyone who knew it. Why would the owner of Galileo's report want to let anyone else in on the secret if this knowledge would give him the edge?

Shakespeare and his alehouse companions who played the game would certainly not have been in on the secret. But even the players who did know the correct odds often found themselves on the losing side, because unscrupulous opponents often used loaded dice.

Fake and loaded dice were a widespread problem. If you roll two dice then the most likely outcome is a total of 7, and the least likely scores are 2 and 12. But if one of the dice has been fixed so that it nearly always lands on a six, the odds of different outcomes for two dice are changed radically. A total of 7 is now no more likely than 12, and if your opponent (the owner of the dice) knows this, they are going to quickly clean you out.

Shakespeare refers to loaded dice more than once, most famously in *Much Ado About Nothing*, when Don Pedro says to Beatrice:

Come, lady, come, you have lost the heart of Signior Benedick.

To which Beatrice replies that Benedick had previously won her heart dishonestly, or in her words:

He won it from me with false dice.

False dice seem to have been disturbingly common. The Museum of London has about 20 dice in its collection, and several of them have been deliberately tampered with.

There were two classic ways to bias a die. The easiest was to duplicate numbers: instead of numbering a die 1 2 3 4 5 6, you might number it 4 5 6 4 5 6, so that no score less than eight would be possible. More subtle would be to simply replace one number with the one on the opposite face. Since the opposite sides of a die always add to 7, you could replace the 1 with a 6, so that if a 6 came up on top of the die, its duplicate was hidden from view on the bottom.

More subtle still was to weight the dice. This was done by drilling a small hole into one of the dots on the number that you wanted to land on the bottom and filling it with lead. More devious tamperers even used liquid mercury. A firm tap on the table could jolt the blob of mercury from one internal chamber to another, shifting its bias. With sleight of hand, these fake dice would be introduced partway through a game, and then secretly replaced with real dice for inspection after the game.

But cheats beware. Dice 'cogging',[14] as it was known, wasn't just illegal. If you were caught, you might be subjected to a public flogging or worse – quite a hazard in itself.

The two-dice version of Hazard lived on for several hundred years. A low score of 1–1 or 1–2 on the dice was known as a 'crab', and this word would eventually evolve to become 'craps', the American name for Hazard which remains popular to this day.

---

14 In *Love's Labour's Lost*, Biron is flirting with the masked Princess, rolling imaginary dice to earn him the chance to speak to her. 'Since you can cog, I'll play no more with you!' she says, fobbing him off. This is a bit rich, given she is in disguise and therefore cogging him already.

Why 10 is a better bet

The consensus among mathematicians in the sixteenth century seems to have been that, when rolling three dice, you were equally likely to get a score of 9 or 10. Their reasoning was to look at the different ways in which three dice can add to those totals. There are six combinations that will score 9 and six that will score 10, like this:

Rolls that add to 9   Rolls that add to 10

Rolls that add to 9	Rolls that add to 10
1, 2, 6	1, 3, 6
1, 3, 5	1, 4, 5
1, 4, 4	2, 2, 6
2, 2, 5	2, 3, 5
2, 3, 4	2, 4, 4
3, 3, 3	3, 3, 4

It was Galileo who spotted the flaw in this argument. Yes, you can score 9 by rolling a 1, 2 and 6, but if you think of the three dice as being different colours, red, white and blue for example, then the numbers could appear on the dice in a different order. They could be 1 (red), 2 (white) and 6 (blue), or 1 (red), 6 (white) and 2 (blue), or 216 or 261 or 612 or 621. In all there are six different permutations. For 1, 4, 4 there are only three permutations, 144, 414 and 441, while 3, 3, 3 can only come up in one way. In total there are 25 ways of getting a total of 9 but there are 27 ways of scoring 10. In other words, you are almost 10% more likely to score 10 than 9. It's not much of a margin and would not easily be spotted over a short number of rolls, but in the long term, betting on 10 would give you a better return. Experienced Elizabethan dice players would have sensed this, but they didn't have the math to prove it.

NINE MEN'S MORRIS

The Nine Men's Morris is fill'd up with mud.

A MIDSUMMER NIGHT'S DREAM

Another Elizabethan recreation was Nine Men's Morris. It's an ancient game, and the word 'morris' is probably a corruption of the Latin word *merellus*, which means a game piece. It was a game of skill rather than chance, played with counters on a board like this:

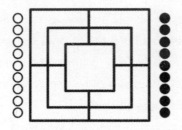

Each player starts with nine counters off the board, and takes it in turns to place a counter on one of the junctions on the board. The idea is to get three of your own counters in a straight vertical or horizontal line, at which point you can remove an opponent's piece. If you can reduce your opponent to just two pieces, you win.

Here's an example of a game in play. White has just formed a line of three and can therefore remove one of the black pieces:

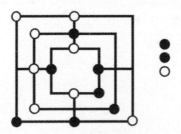

The game would usually be played by two people on a small board. Boards were often improvised. On a visit to Portsmouth to see the artefacts from Henry VIII's ill-fated flagship *The Mary Rose*, I was surprised to see a morris board that had been crudely carved into the top of a barrel. It's unlikely that the sailors got a chance to play during that ship's brief voyage. Other boards have been found scratched onto church pews – a handy game to play during those dull sermons, perhaps?

Nine Men's Morris could also be played on a larger scale. In *A Midsummer Night's Dream*, Titania bemoans the miserable weather so far that summer. When she observes that 'the nine men's morris is filled up with mud' she is referring to an outdoor version that was played by two teams of people, the black team and white team (say). The 'board' markings had been carved into the ground. Players from each team would take it in turn to move to a node (corner) as if they were counters. It was a fun team game to play on sunny afternoons, but rather less fun in muddy conditions.

There is a simpler version of this game called Three Men's Morris that was at least as popular in Shakespeare's day.

This time each player has just three counters, and takes it in turns to place a counter on the board. Again the objective is to get three in a line, but this time that's the end of the game. The first player is not allowed to place their counter on the central spot, as that would mean they could force a certain win.[15] After that, players can place counters

---

15 There's an explanation in the appendix, page 185, of how to win at Three Men's Morris.

anywhere they want. If there isn't a line of three after all six pieces have been placed, the players take it in turns to move a piece to an adjacent node until one manages to form a line.

It's a deceptively tactical game and, although expert players can always avoid defeat, it's easy to make a slip and leave an opening for your opponent to grab victory.

Although Shakespeare doesn't mention this game, he would have known of it. It's quicker to set up than Nine Men, and it takes less time to play. Does it remind you of anything? This was the precursor to the modern game of Noughts and Crosses. Over the next hundred years Three Men seems to have largely died out. As pencils became more commonplace, Noughts and Crosses, later renamed Tic Tac Toe[16] in America, was a more versatile replacement. Tactically, however, Three Men's Morris is a more interesting game. You should give it a try.

---

16 In *Measure for Measure*, Lucio refers to a game of 'tick-tack'. This was similar to backgammon, but its name might have been borrowed for tic-tac-toe centuries later.

NODDY, PRIMERO AND OTHER CARD GAMES

Have I not here the best cards for the game
To win this easy match play'd for a crown?

KING JOHN

Another pastime that saw a surge of popularity in Shakespeare's time was card games. Playing cards were probably invented in China in the tenth century and had been around in England since the early 1400s, but the cards were expensive to make, so in the early days card games were a pastime of the rich. As production quality improved and costs dropped, the popularity of card games exploded. In most countries the idea of four suits with thirteen cards each was widely established, but the suits varied from country to country. The French chose clubs, diamonds, hearts and spades as their four suits, and since almost all of the packs of cards used in England were manufactured in France, the English adopted the French model, and that has been the norm ever since.

There is no definitive explanation for why there are four suits and 52 cards, but it is perhaps not a coincidence that there are also four seasons and 52 weeks in the year. Even if there was no direct link between the suits and the seasons, the numbers would have been comfortably familiar.

Two card games were particularly popular. Noddy (the ancestor of cribbage) was usually played just for fun. It involved adding the values of cards in your hand, with particular combinations, or sets, earning points. All this adding up made Noddy a handy everyday exercise in mental arithmetic. The winner was the player who first reached 31 or more points. The word Noddy meant a fool, and referred to the knave card (known as the knave noddy) which had a special value in the game.

The game is mentioned in *Two Gentlemen of Verona*, but only as an excuse for a pun that makes fun of one of the characters. Proteus has asked a servant called Speed to deliver a letter to his love, Julia, and wants to know her response.

> Proteus: But what said she?
> Speed: (nods his head). Ay
> Proteus: Nod ~ Ay ~ why that's noddy! [i.e. 'you're a fool, Speed']
> Speed: You mistook sir, I say she did nod, and you ask me if she did nod and I say 'Ay'.
> Proteus: And that set together is noddy.

The audience would have known that in the game of noddy the idea was to make sets. It would be interesting to know if this pun raised a laugh.

The other major card game of the time was called Primero. The game was a little like modern poker. Players were dealt four cards, and then wagered over which player had the best hand. A hand in which each of your four cards had a different suit was called a Primero, and would usually be good enough to win. However an even better hand was a Fluxus (all four cards of the same suit, which we now call a 'flush'), and best of all was a Chorus which was four of a kind. The fact that Primero was played for money made it extremely popular. Like other gambling games, however, it could lead players to ruin. In *The Merry Wives of Windsor*, Falstaff declares:

> I never prospered since I forswore myself at primero.

This is a confession that he cheated at the game,[17] though it's not clear if it also means that he lost money as a result, or just lost friends.

---

17 That's what most modern academic interpretations say, at least.

For those prone to the slippery slope of immoral living, including gambling, there was wise advice from *King Lear*'s Fool:

> Have more than thou showest,
> Speak less than thou knowest,
> Lend less than thou owest,
> Ride more than thou goest,
> Learn more than thou trowest,
> Set less than thou throwest . . .

'Set less than thou throwest' means 'don't bet everything on the next throw of the dice' – sound tactics for any gambler.

But such sound advice was not always heeded. And, needless to say, all this fun, often laced with heavy doses of immorality, did not go down well with the Church.

It would all end in tears. Just thirty years after Shakespeare's death, Oliver Cromwell and his fellow Puritans deposed the monarchy. One of their first acts was to ban all forms of theatre, sport and the playing of games, and the recreational math that went with them.

CHAPTER IV
MONEY

SHAKESPEAREAN CURRENCY

*I give thee this pennyworth of sugar, clapped even now
into my hand by an underskinker, who never spoke other
English in his life than 'Eight shillings and sixpence'*

HENRY IV PART 1

Money runs as a rich seam throughout Shakespeare's work: everywhere you look there are people who owe money, or are trying to steal it, pay it, borrow it, lend it or earn it. And, of course, strong numeracy is fundamental if you are going to deal with money. For anyone conducting business in the Shakespearean era it would be impossible to avoid dealing with fractions on a daily basis.

In sixteenth-century England there were no banks,[18] and no bank notes or cheques, so with the exception of the occasional credit note (a sort of IOU), all money was stored and transacted in the form of coins.

---

18 It would be several decades before the English caught up with the banking innovations of Italy, Germany and other European countries.

As we have seen, English society had just adopted the Arabic system for recording numbers (a 'decimal' system, because it counts in tens, hundreds, etc.). However, the English currency was *duodecimal*, in other words it was built on twelves, for the practical reason that twelve can be equally divided into halves, thirds or quarters, while with ten you can only divide it into halves and fifths.

The currency was based around the units of pounds, shillings and pence, a system that was to continue in Britain until 1971. Instead of 100 pennies in a pound as there are now, there were 240. The breakdown was:

12 pennies in a shilling

20 shillings in a pound

The actual coinage, however, was a weird and wonderful collection of fractions and multiples of those basic units – and Shakespeare at one point or another referred to nearly every type of coin that was in circulation. Almost all the coins in everyday use were made of silver, though there were some prestigious high-value gold coins, too. Starting with the coins of lowest value, the table opposite shows the ones that existed in England, and the number of mentions they get in Shakespeare's work.

What an eccentric collection it was. This diverse range of coins meant there were numerous different ways of conveying a particular amount of money. For example, twelve pennies, three groats and two sixpences all had the same value of one shilling.

An Elizabethan sixpence.

Coin	Value	No. of mentions	Comments
Farthing	¼ penny	9	The lowest value of coin, but still made of silver.
Halfpenny	½ penny	7	Usually pronounced 'hay-pny' and written 'ha'penny'. Also called an obolus. In *Henry VI Part 1*, Falstaff's bread is listed as costing 'ob' (half a penny).
Three-farthing	¾ penny	3	A coin introduced by Elizabeth I. When Shakespeare mentions three-farthing he might be referring to the single coin, or simply three of the smaller value coins.
Penny	1 penny	25	Written as 1d, the 'd' standing for 'denarius', an ancient Roman coin.
Twopence	2 pennies	1	Usually pronounced 'tuppence'.
Threepence	3 pennies	0	Pronounced 'threppence'. In later times the coin was often referred to as a 'thrupenny bit'
Groat	4 pennies	11	Also occasionally called fourpence, but never by Shakespeare.
Sixpence	6 pennies	10	Also called a tanner (by the Clown in *Hamlet*), a tester (by Falstaff in *Henry IV Part 2*) and a testril (by Sir Andrew in *Twelfth Night*).
Shilling	12 pennies	15	Written as 1s. Note that the 's' stands for 'solidus', another Roman coin, not for shilling.
Half crown	2½ shillings	0	Half crowns were popular from their first minting by Henry VIII in 1526 until the late 1960s, but never mentioned by Shakespeare.
Crown	5 shillings	30+	Shakespeare's most popular coin. Sometimes it isn't clear whether he is referring to the coin or the royal headgear – and sometimes that is deliberate.
Noble	6s 8d (one third of a pound)	4	In pre-decimal currency it was handy that a pound could be divided exactly into thirds in this way (that was the benefit of having 12 pennies to the shilling).
Royal	10 shillings	1+	The royal was an upgrade to the noble.
Angel	10 shillings (or so)	1	Mentioned in *The Merchant of Venice*: 'They have in England a coin that bears the figure of an angel stamped in gold.' The value of the coin changed several times as gold became more valuable compared to silver. An Edward VI angel was worth 10 shillings, but a James I angel was worth 11 shillings.
Pound	20 shillings	25+	Until 1583, a pound was only a value used in accounting and wasn't actually a coin. Elizabeth was the first to mint pound coins, but they were relatively uncommon.
Sovereign	20 or 30 shillings	0	Sovereigns were first minted by Henry VII and later by James I of England but were intended more for prestige than for payment. Henry VIII also minted half-sovereigns.

Shakespeare indulged himself in thinking up a multitude of ways to express money values. When Falstaff asks his page how much cash there is in his purse, Shakespeare could have had the page reply that there were two shillings and sixpence, or thirty pennies. Instead he replied

Seven groats and two pence

which is the same amount – but it takes a little calculation to prove it.

Or take this little exchange from *Love's Labour's Lost*:

Biron: What is a remuneration?
Costard: Marry, sir, halfpenny farthing
Biron: Why then, three-farthing worth of silk.

Since a farthing is a quarter of a penny, and a halfpenny is worth two farthings, Costard's halfpenny farthing is the same as Biron's three-farthing. This was possibly the first time in English literature that there had been a bit of casual banter based on the addition of fractions. And possibly the last.

The names of coins also lent themselves to punning. With stories full of kings and queens, and coins called nobles, crowns and royals, there were plenty of opportunities for word play.

In *Henry IV Part 1*, Mistress Quickly comes to report a visitor to Prince Henry, and Henry asks her to pay him to go away:

Mistress Quickly: Marry, my lord, there is a <u>noble</u>man of the court at
*   door would speak with you: he says he comes from*
*   your father.*
Prince Henry: Give him as much as will make him a <u>royal</u> man,
*   and send him back again to my mother.*

Henry deliberately uses a double meaning. A 'noble' (6s 8d) man becomes a 'royal' (10s) man if you pay him 3s 4d. Geddit? (This was an anachronism by the way, the special 'royal' coins didn't appear until long after the reign of Henry V.)

As if that list of coins wasn't enough to choose from, Shakespeare also brought in numerous international coins as well, though at times the choice seems a little random. Foreign coins that he features include Dutch guilders (in *The Comedy of Errors*, set in what is now Turkey), Italian chequins (in *Pericles*, set in Turkey and Lebanon), Portuguese crusados (in *Othello*, Venice) and Greek drachmas (in *Julius Caesar*, Italy).

It's fair enough to bring in foreign money – most of the plays are set in continental Europe – but he could at least have chosen money that belonged to the country the play was set in. He gets away with Italian ducats in *Hamlet*'s Denmark since they were legal tender in many countries, as euros are today.

In *Timon of Athens*, Shakespeare chose talents as the unit of money, though as with all his other international units of money, he was pretty loose with his exchange rates. Talents were not a regular currency, but a single talent could be the equivalent of 50 pounds (weight) of silver. In one scene, a messenger tells Lucilius that Timon needs to borrow fifty-five hundred (5500) talents, which is the equivalent of about 120 tonnes of silver – worth about £85 million in today's money. That's an eye-wateringly implausible amount to be asking your friend to lend you. It's possible that Shakespeare wanted to make a point here that Timon was asking for an impossible sum and was clearly doomed – but it's just as likely that Shakespeare plucked a number out of the air that sounded big, without caring too much about what a talent was actually worth.

As well as going big with money, Shakespeare sometimes goes very small. Several times he refers to coins that are worth next to nothing to make a point. When Falstaff says 'I'll not pay a denier' he's referring to an old French coin that was worth about one tenth of a penny. That's as close to zero as Shakespeare gets without saying it.

The word 'money' itself appears in thirty of Shakespeare's plays, most frequently in *The Merchant of Venice* (which is not surprising) and in *The Merry Wives of Windsor* (which seemed surprising to me until I discovered that the plot centres on the character Falstaff, who is looking for ways to solve his financial woes).

One of the few plays in which the word money is *not* mentioned is the Scottish play, *Macbeth,* though in one scene Ross talks of demanding ten thousand 'dollars' from the King of Norway. Quite why he is talking about dollars in the eleventh century is another matter – the word 'dollar' didn't appear until about 1600; it was the English word for Spanish 'pieces of eight' coins. Shakespeare had probably heard the word for the first time not long before writing *Macbeth*, and thought it would be fun to use it.

60	*The First Part of K*

He searcheth his Pockets, and findeth certaine Papers.
Prince. What haft thou found?
Peto. Nothing but Papers, my Lord.
Prince. Let's fee, what be they? reade them.

Peto. Item, a Capon.	ii.s.ii.d.
Item, Sawce.	iiii.d.
Item, Sacke, two Gallons.	v.s.viii.d.
Item, Anchoues and Sacke after Supper.	ii.s.vi.d.
Item, Bread.	ob.

This extract from Henry IV Part 1 *(first folio) gives an indication of typical prices. A capon (well-fed chicken) is 2 shillings and tuppence. Bread is 'ob' meaning a ha'penny.*

What was money worth?

It's hard to pin down what prices and incomes were in Shakespeare's time; different sources give a wide range of figures, but here is a guide to what you might have earned and what everyday items might have cost in the year 1600. As a reminder, pounds, shillings and pence are written as £ s d.

Occupation	Incomes	
	Per week	Per year
Labourer	2s (= 24d)	£5
Skilled craftsman	5s (= 60d)	£12
Vicar or schoolmaster	8s (= 96d)	£20
Tradesman	8s	£20
Country gentleman	£2	£100
William Shakespeare	£2	£100ᵃ
Merchant	£5	£250
Nobleman	£400	£20,000

Everyday items	Prices
Loaf of bread	2d
Best beef	3d per pound
One chicken	1-2s
Raisins	3d per pound
Pint of ale	1d
Six sheets of paperᵇ	1d
Cheapest Globe seat	1d
Globe seat with cushion	6d+
Lodging at inn	2d per week
Small book	1s
Shirt	1s
Pair of shoes	1s
Gardening gloves	10d
Decorative gloves	£1+

ᵃ This is an educated guess based on the earnings of his peers, and it would have varied significantly from year to year. In 1600, Shakespeare might have earned, say, £25 for selling his plays and £75 from his share of Globe receipts. At his peak in the early 1600s, some have speculated that Shakespeare earned as much as £200 per year, but in years where the plague struck and theatres were closed, his income would have been a fraction of that.

ᵇ The idea that paper was a precious commodity was challenged by Heather Wolfe of the Folger Shakespeare Library. Paper was certainly very expensive compared to the modern day, but even a labourer could have afforded to buy a few sheets (in the unlikely event of ever needing to use them).

SHAKESPEARE THE ACCOUNTANT?

Against that time, if ever that time come,
When I shall see thee frown on my defects,
When as thy love hath cast his utmost sum,
Called to that audit by advis'd respects;

SONNET 49

While researching the links between Shakespeare and money, I was surprised to come across a 1974 article by Bernard Reynolds of the Cardiff College of Commerce entitled 'Was Shakespeare an Accountant?'. His suggestion was that Shakespeare's language and metaphors are so frequently drawn from accounting that he may have been well versed in the profession himself. At first I thought it was just a tongue-in-cheek paper, but looking through the many quotations that he referenced, I began to realize that he had a point.

Here for example is a quote from *Cymbeline*, in which the Jailer likens the hangman's noose (a rope or cord that can be bought for just a penny) to – of all things – an accountant sorting out the debits and credits of the victim's life:

O, the charity of a penny cord! it sums up
thousands in a trice: you have no true debitor and
creditor but it: of what's past, is, and to come, the
discharge: your neck, sir, is pen, book, and counters;
so the acquittance follows.

A discharge is an accounting statement that releases somebody from debts. The 'counters' are referring to the common method by which accounts were added up – using coin-like tokens called jettons on a lined counting board, similar to an abacus. (Most accountants were more confident adding and subtracting using counters than using the new arithmetic methods, however efficient the latter were.) For many years Shakespeare was a shareholder in his own theatre. He also bought and sold property from which he earned rent. He would have had to manage his own quite complex personal finances, and perhaps he helped to look after the books of his company as well (somebody had to!). And he was able to retire to Stratford, where he died in 1616 a relatively wealthy man.

The common term for doing money calculations and sorting out accounts was 'reckoning', and across all his published work Shakespeare uses reckon or reckoning over forty times, though often the word is being used metaphorically. In *Macbeth*, when Malcolm becomes king and says to his soldiers:

> We shall not spend a large expense of time
> Before we reckon with your several loves
> And make us even with you

the debts he is paying off are ones of loyalty and friendship, rather than money.

There are other clues that Shakespeare had more than a passing acquaintance with the world of trade and accounting. In several plays he refers to 'marks', for example in *The Comedy of Errors*:

> Where is the thousand marks I gave thee, villain?

In England, a mark was a monetary unit equivalent to 160 pennies (two thirds of a pound), but it was never a minted coin. A mark was a term used in bookkeeping when logging transactions. If a merchant sold an item for, say, 26 shillings and 8 pence – an amount equal to 320 pennies – a bookkeeper might have written 'sold for two marks'.

Shakespeare could just as easily have had his character demand the 'thousand pounds' or even the 'four thousand crowns I gave thee, villain'. But instead he brought in an accounting term that would probably have been less familiar to his audience.

There is also evidence that Shakespeare was familiar with the tax rates that were in operation. There was a tax that had been in place since 1334 called the 'fifteenth'. It required citizens with wealth or income above a certain base threshold to pay one fifteenth of the value of their 'movable property' as a tax. In addition, there was another tax (known euphemistically as a 'subsidy') that was sometimes imposed if the monarch was in need of cash, for example to help fund an army against a Spanish armada.

Tucked away in *Henry VI*, a messenger announces the arrival of Lord Say, who has a reputation for charging excess taxes:

My lord . . . here's the Lord Say . . . he that made us pay one and twenty fifteens, and one shilling to the pound, the last subsidy.

The messenger mentions both types of tax here. First he talks about a tax of twenty-one fifteen(th)s. Taken literally, this means a tax rate of 21/15, an extortionate 140%, which is more than the goods being taxed are worth! Tax historians regard this as an example of Shakespearean exaggeration, to indicate how unfair the people thought these taxes were. On top of the fifteenths, is the 'subsidy' tax of one shilling in the pound, which is a 5% levy. A tax of exactly this amount had been imposed by Queen Elizabeth in 1589, so it seems that Shakespeare was making a topical reference to an unpopular tax that richer members of his audience would have been only too aware of.

It should also be said that Shakespeare was not fond of paying taxes himself. For example, in a London tax commissioner's report of 1597 for St Ellen's parish, a 'William Shackspeare' (Elizabethans were very flexible with spelling names) is listed as owing five pounds and five shillings, the equivalent of perhaps £1000 today.

For all his references to bookkeeping, Shakespeare does give the impression via his characters that, while he was good at handling numbers, it was not a skill he was particularly proud of. In *Henry IV Part 2*, Falstaff suggests that the only skill for which intelligence is useful is working out the bill in a tavern – but what a waste of a smart brain:

. . . a [barman's] quick wit wasted in giving reckonings, all other gifts . . . are not worth a gooseberry.

Meanwhile, do you remember Bianca, who couldn't bear the thought of waiting a week ('eight score eight hours') till she could see her lover again? She sums up her frustration by saying how it will feel counting the sundial as it goes through the hours 160 times.

More tedious than the dial eight score times?
O weary reckoning!

'Doing accounts is a chore', she is saying.

It has been joked that the bust of Shakespeare above his tomb made him look like 'a self-satisfied pork butcher'.[19] But when I visited Holy Trinity Church in Stratford to look at the famous effigy, it suddenly dawned on me that the face in front of me was more like that of my accountant than my local butcher. That article in 1974 was onto something.

---

19 First said by the Shakespeare scholar John Dover Wilson.

BORROWING, LENDING
AND INTEREST

Neither a borrower nor a lender be;
For loan oft loses both itself and friend,

HAMLET

O ne of the most intriguing aspects of finance in Shakespeare is the thorny issue of lending and borrowing money. Money-lending was an essential part of the economy in Elizabethan times. It would have been almost impossible to conduct business without it. Cash was needed to purchase everything from chickens to houses, and if you didn't have the money to hand, you would have little choice but to resort to borrowing.

Unfortunately, lending money and charging interest for it was severely frowned upon. This practice, known as usury, was deemed un-Christian, and for centuries it had therefore become one of the few occupations that Jews were able to pursue, having been excluded from many other professions.

In *The Merchant of Venice*, the merchant Antonio agrees to lend money to his young friend Bassanio at no charge, much to the disgust of the Jewish money-lender Shylock, who thinks that Antonio's reputation as somebody who lends money for free is undermining the profession.

He lends out money gratis and brings down
The rate of usance here with us in Venice.

Antonio was a fictitious character, but there were real people who lent money without interest, particularly to help friends. The diary of Philip Henslowe, who managed the Rose theatre, a rival to Shakespeare's

Globe, reveals that he regularly lent money to actors, playwrights and other associates, apparently with no charge. As was normal for that time, the loans were quite short term. In general, Elizabethan loans were for only a few weeks or months. Long-term loans – such as a mortgage to purchase a house – were almost unheard of.

How did Shakespeare buy his house?

It's interesting to speculate how Shakespeare went about purchasing New Place, the grand house that he acquired in Stratford-upon-Avon in 1597, a few years before his career reached its peak. The house cost him £120. It has been conjectured that this was roughly double his annual income at the time, a reflection both on how well paid he was even when he hadn't reached his peak earnings, but also how relatively cheap property was compared with today. (If you wanted to buy a detached house for twice your salary in central Stratford today, you'd need to be earning several hundred thousand pounds a year.) Shakespeare would have had to pay for this in cash, perhaps in more than one instalment, but where did the cash come from? Without banks or safe-houses to store it, he must literally have saved up coins, perhaps kept in a locked case under a floorboard, and when the time came, carried this money with him to Stratford to hand it over. £120 would most likely have been kept in crowns worth five shillings each. £120 would be 480 crowns, each one weighing about an ounce (30 grams). In total that's about 14 kilograms of silver coin to lug on horseback from London, too much to hide around his body away from the sight of highwaymen. Perhaps he took his money back in smaller purses over several journeys, and kept it stored in Stratford.

However, everyone knew that lending without getting any return carried a risk, and even if the loan was repaid in full, by the time the lender got their money back it would have lost purchasing power if prices in the market were rising – and inflation was an issue across Europe, thanks to the influx of huge amounts of silver from South America in particular. What often happened, therefore, is that lenders would find ways to make money out of lending without declaring an interest rate. If they were lending to somebody who was in trade, they might agree to take a cut of the profits. For those who traded internationally, there was an even simpler scam: they would simply fudge the exchange rate. A loan of £200 to a merchant might be made at an exchange rate of 11 guilders per pound (so 2200 guilders), but when that 2200 guilders was paid back, the agreed exchange rate had miraculously moved to 10 guilders per pound, meaning the lender now had £220, a 10% gain. Few Christian lenders were as virtuous as Antonio or Henslowe (sometimes) appear to have been, even if they pretended to be.

How could society turn a blind eye to this 'necessary evil'? The answer was a bit of political gaslighting from Queen Elizabeth. In 1571 she introduced 'An Act Against Usury', in which nobody could lend at an interest rate higher than 10% (or as her act put it: 'Any rate above ten pound for the hundred per year shall be void'). While the Act claimed to be anti-usury, it was actually making usury legal, so long as the charges weren't too extortionate.

The Merchant of Venice isn't the only play where Shakespeare portrays usurers as contemptible. In *Henry VI* the Duke of Gloucester describes the Bishop of Winchester as a 'pernicious usurer', while in *Cymbeline* usurers are described as 'vile men' for charging a third (33%), a sixth (16%) or a tenth (10%).

I know you are more clement than vile men,
Who of their broken debtors take a third
A sixth, a tenth . . .[20]

Taking a tenth might be deemed vile, but Queen Elizabeth had made it legal, and presumably one twentieth (5%) would have been acceptable.

Throughout his plays, Shakespeare makes clear that he is not on the side of the money-lenders who charge interest. What's curious, however, is that his own father John was accused not once but twice of usury, in 1570. On both occasions he had allegedly charged interest of 20–25%, much higher than the 10% that Queen Elizabeth's Usury Act would make legal the following year.

John Shakespeare was never found guilty, but he did make a settlement for one of the cases out of court to clear his name, which suggests that he might have had something to hide. Whatever the truth, John Shakespeare's previous position as High Bailiff of Stratford seems to have helped him. Despite his dodgy business dealings, he was still listed as one of the 'gentlemen of Warwickshire' in a government list of 1580.

---

20 It is a classic rhetorical device for items in a list to start small and get bigger. The numbers 3, 6 and 10 follow this pattern but, as fractions, one third, one sixth and one tenth get progressively smaller.

CHAPTER V

MEASURE FOR MEASURE

AN ERA OF IMPRECISE MEASUREMENTS

The English lie within fifteen hundred paces of your tents.

HENRY V

The sixteenth century was an era of imprecision, at least for the average person in the street. These days we track and measure things with precision partly because it often matters, but also because we can. In Shakespeare's time, counting large numbers didn't matter so much, and when it came to measurement, in everyday life few things needed to be measured to a high degree of accuracy – even if they could be. There were no millimetres marked on rulers, no second hand on a watch, and as for temperature, there was merely hot, warm and cold.

A carpenter's rule found on Henry VIII's ship, the Mary Rose. *The 'inches' are marked quite crudely and are slightly smaller than the modern inch. The shorter lines mark the half and quarter inches.*

This imprecision in numbers is reflected in the way Shakespeare deals with numbers that are linked with everyday life. Again and again, he puts hand-wavy numbers into a character's speech, using 'about' or 'odd' to indicate the number's 'ish-ness'. Here are some examples:

About some half hour hence	(*Cymbeline*)
About the ninth hour	(*Julius Caesar*)
Eighty odd years	(*Richard III*)
Of wounds two dozen odd	(*Coriolanus*)

But there's another way to indicate that precision doesn't matter, which is by using two numbers instead of one. For example, we might say: 'Just give me *two or three* minutes and I'll be with you.'

Shakespeare uses this 'X or Y' form of imprecision all over the place, in fact he does it so often that I was able to collect examples for every consecutive pair of numbers from 1 up to 10, with plenty of spares (see the box on the next page).

The exception to all of this was weights. When buying produce from the market, or more valuable materials such as metals, knowing the weight accurately was important if you weren't to be ripped off.

Weights could be measured to the nearest grain, as in a single grain of wheat. That's what is being referred to in *Troilus and Cressida*, when Ulysses talks of 'every grain of Plutus' gold'. Other obsolete

Who cares about the exact number?

1-2   I pray you, tarry, pause <u>a day or two</u> [*The Merchant of Venice*]
or [He] swears a prayer or two   [*Romeo and Juliet*]

2-3   Tis two or three my Lord that bring you word that Macduff
is fled to England   [*Macbeth*]

3-4   I was forced to wheel three or four miles about [*Coriolanus*]

4-5   There's four or five to great St Jaques bound   [*All's Well That
Ends Well*]

5-6   Five or six honest wives that were present   [*The Winter's Tale*]

6-7   Six or seven more winters of respect   [*Measure for Measure*]

7-8   Stage Direction:* Enter Seven or eight citizens [*Coriolanus*]

8-9   I will last you some eight year or nine year   [*Hamlet*]

9-10 Nine or ten times   [*Othello*]

* To be fair, this stage direction might have been written by someone other
than Shakespeare.

small weights that Shakespeare refers to several times are the 'scruple' (which was officially the weight of twenty grains) and the 'dram' (three scruples). There were eight drams in an ounce and then – depending on what you were weighing – either twelve or sixteen ounces in a pound.[21]

In *The Merchant of Venice*, there is a reference to this weighing system. In the play, Shylock has famously demanded a 'pound of flesh' instead of charging interest on his loan. Portia says that this measurement has to be precise, in fact any deviation from one pound should be within

the twentieth part of one poor scruple.

---

21 The Troy and Apothecary pounds were used to measure jewellery and medicines; these
had 12 ounces. For everything else, such as wool and apples, there was the Avoirdupois
pound, which had 16 ounces.

Since one twentieth of a scruple is a grain, she's saying that Shylock's cut has to be precise to within one grain, or 60 milligrams, of the exact amount. Any more than that and he'll be sentenced to death.

The word 'scruple' originally meant a tiny stone. The modern sense of 'having scruples', meaning having a conscience, comes from the idea of having a tiny stone in your shoe that makes you feel uncomfortable.

The fact that numbers in Shakespeare's time were usually so imprecise means that when he *did* state a number precisely, his audience would really have noticed it. In *Troilus and Cressida* 'sixty and nine' (69) royal rulers are reported to have set out to ransack Troy, not sixty-eight or seventy. The point here is that these are sixty-nine important individuals, no more no less, each with a desire for revenge.

This was not the only time that Shakesepeare used precise numbers to reflect the status of the people being counted. In *Henry V*, the king gets a note to inform him that ten thousand French soldiers have been slain (a conveniently rounded number), of whom precisely 'one hundred twenty six' are nobles. That's a round number for the commoners, but a precise number for the nobility; Shakespeare could just as easily have said that the nobles numbered 'a hundred-odd'.

LENGTH AND DISTANCE

How many inches doth fill up one mile?

LOVE'S LABOUR'S LOST

O f all the length and distance measurements in Shakespeare, the mile is the most common. Today we know exactly how long a mile is thanks to international statutes – it is 1760 yards, 5280 feet or 1609.344 metres, wherever you are in the world. However, in the sixteenth century it was a more flexible measure. It could vary by 25% or more, depending on whether it was an 'old London' mile (5000 feet), a Scottish mile (5952 feet) or an Irish mile (a whopping 6720 feet). There were local variations too: a mile in Hampshire might be different from a mile in Dorset.[22] And, overseas, the range was even wider. The most precise that Shakespeare ever gets with distance is in *The Winter's Tale* when Autolycus says that he has

a kinsman three quarters of a mile hence.

But since *The Winter's Tale* is set in both Sicilia and Bohemia, which mile does he mean? An Italian mile was around 1600 metres, whereas in central Europe a mile could be four times as long as that, which is quite a difference. Though, to be honest, it's unlikely that Shakespeare or his audience would have cared much. Nobody was measuring.

Queen Elizabeth cared, however. As travel became more important, it was time to agree on an official length of a mile, and in 1593 she put through the Weights and Measures Act of Parliament, which established the length of the English statute mile. Up until then, there had been two ways of measuring a mile, one based on the length of

---

22 As late as 1677, in Robert Plot's *The Natural History of Oxfordshire*, he talks about there being three sorts of mile in the county, the 'greater', the 'lesser' and the 'middle' mile, and maps might show all three scales.

a ploughed strip of land (the furlong), the other based on the length of a foot. These two had once tallied with each other, but the official length of a foot had been reduced in the thirteenth century because (it is claimed) King Henry redefined the yard, which is 3 feet, as being the length of his own arm. This had led to a divergence in the two ways of measuring the mile, depending on whether it was measured by yards or by furlongs.

It is thanks to Elizabeth I that we have 1760 yards in a mile. And by the time Shakespeare wrote *Love's Labour's Lost* in 1596, there was an official answer to Boyet's question 'How many inches doth fill up one mile?'. He wasn't expecting an answer, but for the record, a mile is 63,360 inches.

The system for measuring length was a curious collection of units, and, just as with money, it was all based on a duodecimal system. The table on the next page shows the commonly used measures in descending order of length, and the frequency with which Shakespeare mentions them.

The list in the table includes a few terms that have disappeared from general use. While Shakespeare's preference was to use miles when talking about long distances, he often used leagues instead. A league is three miles, but it was normally used colloquially, to indicate roughly how far you can walk in one hour. When in *A Midsummer Night's Dream* Lysander says 'a league outside the town, where I did meet thee once', you can picture that this rendezvous took place about an hour's walk away.

Unit of length		No. of mentions	Comments
League	Roughly three miles	13	
Mile*	8 furlongs (1760 yards)	50	
Furlong*	220 yards	2	Originally a 'furrow long', deemed the distance that an ox could pull a plough before taking a rest. Furlongs are still used in horse racing, particularly to mark out the final mile.
Chain	22 yards	0	Used to measure plots of land, still to be found on sports fields across the world: the length of a cricket pitch is one chain and the width of a football penalty area is two chains.
Perch*	5½ yards	1	Also known as a 'rod' or 'pole', neither used by Shakespeare. Four perches make a chain.
Fathom	6 feet	13	Used only to measure depth of water.
Ell	1¼ yards (45 inches)	5	Used by tailors to measure cloth.
Scottish ell	1 yard and an inch (37 inches)	0	'Yard and inch' was a standard measure for cutting cloth. The additional inch was originally added as a tax dodge but became approved practice written into law.
Yard*	3 feet (36 inches)	10	Old English measurement. Queen Elizabeth I established the 'Exchequer yard' which was the official standard for over 250 years.
Foot/feet*	12 inches	5	Based loosely on the length of a adult male foot.
Inch*	3 barleycorns	29	Roughly the length of the top joint of an adult thumb. The word 'inch' comes from the Latin *uncia* meaning one twelfth. (Ounce has the same origin.)
Barleycorn*	Base unit	0	One third of an inch, about 8 mm in metric measurement. Based on the length of a grain of barley. Shoe sizes still increase in increments of one barleycorn, i.e. a size 11 is 1/3 inch longer than a size 10.

*Those marked with an asterisk were the statutory measures mentioned by law in Elizabethan England.

Remember those creative ways in which Shakespeare likes to express numbers (page 10)? In *The Comedy of Errors* Aegeon says that his ship hit a rock before it could get within 'twice five leagues' of the other ship. A bit of multiplication reveals that the distance between the ships was actually $2 \times 5 \times 3 = 30$ miles. So why not just say 'thirty miles'? It has the same number of syllables as 'twice five leagues', and the line didn't need to end with a rhyme, so we have to assume that Shakespeare was simply being playful with numbers and the sounds of words.

On land, ten leagues (thirty miles) seems to have been as far as somebody would typically travel in a day, even if they rode on horseback, and Shakespeare rarely mentions distances any farther than this. Thanks to the horrendous state of the roads, even a coach might not be able to go this far, as Portia explains in *The Merchant of Venice*:

> But come, I'll tell thee all my whole device
> When I am in my coach, which stays for us
> At the park gate; and therefore haste away,
> For we must measure twenty miles to-day.

The only distances mentioned by Shakespeare that are farther than thirty miles are exaggerations. In *The Tempest*, Stephano boasts that after being shipwrecked he swam 'five and thirty leagues' to the shore (105 miles) – a feat that would have beaten the Guinness World Record even in the twentieth century. And in *Henry IV*, Glendower talks of musicians that are a thousand leagues (three thousand miles) up in the air.[23]

---

23 The most famous use of the word 'league' is in the title of Jules Verne's classic Victorian adventure *Twenty Thousand Leagues Under the Sea*. This refers to the distance the submarine travels and not (as I'd always assumed) the depth. It would be an impossible depth – roughly twice the circumference of the earth.

The other archaic measurement that Shakespeare used more than once is the ell. This ancient measurement was originally based on the length of an adult male arm and it was adopted by tailors as a standard unit of measurement of cloth. The word 'ell' itself has disappeared, but it lives on in the word 'elbow', which means literally the bend (bow) in the arm.

In *The Comedy of Errors*, the servant Dromio complains that he has been ensnared by an overweight woman by the name of Nell. He makes a lot of uncomplimentary remarks about her appearance and when asked her name he replies:

> Nell sir, but her name and three quarters ~ that's an ell and three
> Quarters ~ will not measure from hip to hip.

First there's some mathematical wordplay – 'ell' is three quarters of the word 'Nell' – and then the revelation that Nell is more than 1¾ ells wide, which works out at 6 feet 6¾ inches (just over 2 metres).

TELLING THE TIME

I pray you, what is't o'clock?

Plays at the Globe would typically begin at two o'clock. Without lanterns to illuminate the stage (too much of a fire risk with all that wood and thatch around), there was no choice but to perform in daylight, and in winter the light would be fading at four o'clock.

But if members of the public were to turn up for the start of the show, they needed to know what time it was. How did they know it was two o'clock?

By far the most accurate timekeeper was the sun. If you knew which way was south, then it was a simple matter to know when it was noon. Alternatively, rather than squint at the sky, you could let a needle cast a shadow. Sundials had been around for centuries, and were to be found on walls, and even in people's pockets – miniature pocket sundials were quite a thing.

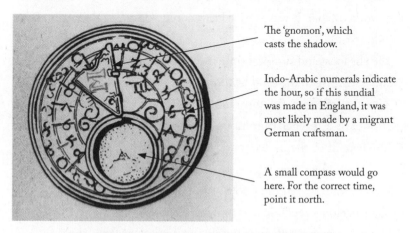

The 'gnomon', which casts the shadow.

Indo-Arabic numerals indicate the hour, so if this sundial was made in England, it was most likely made by a migrant German craftsman.

A small compass would go here. For the correct time, point it north.

A pocket sundial from the mid-sixteenth century.

Shakespeare mentions sundials several times. In *As You Like It*, Jaques describes a fool he has met for whom life seems to be little more than watching the hours go by:

And then he drew a dial from his poke [pocket] . . . and says very wisely 'It is ten o'clock.'

In *All's Well That Ends Well*, Lafeu discovers he has misjudged somebody, and says

Then my dial goes not true: I took this lark for a bunting.

'My dial goes not true' shows just how surprised he is, because the whole point of dials is that they are *always* true, as long as you use them in roughly the same latitude that they have been made for. Two identical sundials, one used in Edinburgh and the other in Madrid, would show noon at the same time because they are roughly the same longitude (about 3 degrees west of the Meridian), but they would disagree on when it was three o'clock because of their different latitude, so the sun would be at a different elevation in the sky.

There was, however, a rather bigger problem with sundials. They only worked when it was sunny. How could you tell the time at night, or indoors?

The cheapest and simplest time measurer for all times of day was the hourglass, in which the sand would take an hour, or – more often – half an hour, to fall from the top bulb to the bottom. A priest delivering a sermon, or a teacher running a class, would often use an hourglass for the benefit of themselves and/or their audiences in knowing when things were going to end. (It is said that some priests would adjust the amount of sand so that their hours would run 'long' or 'short' depending on their personal preference for giving sermons.)

Hourglasses were effective timekeepers at sea as well. The rolling of the ship didn't affect the rate at which sand passed through the hole. In *All's Well That Ends Well* this is referred to as the 'pilot's glass'.

Hourglasses could be used to measure the entire day if you wanted, so long as there was somebody who dutifully turned over the glass every time that it ran out. On ships there would often be a cabin boy who did that – fortunately there were four-hour glasses, offering an opportunity to nap in between turnings. In *The Tempest*, the idea of telling the time this way is suggested in a conversation between Prospero and Ariel:

> Prospero: What is the time of the day?
> Ariel: Past the mid-season. [noon]
> Prospero: At least two glasses.

'Two glasses' of course means two hours here, not that they've just consumed two pints of ale.

Unfortunately, hourglasses were not particularly reliable. They weren't perfectly airtight, so they would let in moisture and eventually the sand would begin to stick. What was really needed for reliable timekeeping was an automatic device that would tell the time for you. And in Shakespeare's world, the clock was coming into its own.

Mechanical devices, powered by a falling weight that would drive a series of cogs, had been around for three hundred years, probably originating with Muslim inventors in the early stages of the Ottoman Empire. The original meaning of 'clock' was a bell and, even in Shakespeare's time, when he mentions a clock you should usually think of it as a mechanism that causes a church bell to be struck every hour. There would be nothing to see – it was almost unheard of to have a clock face with hands showing the time.

Most clocks were in church towers, and most commonly they struck only on the hour. Almost certainly the crowds flocking to the Globe were guided by the sound of two bongs ringing from a church nearby. There were several churches within earshot of the Globe in Southwark, including St Mary Overy (now Southwark Cathedral), and St Paul's Cathedral just across the river. Their clocks weren't

The Salisbury Cathedral clock, from c.1390. It was designed to strike a bell, not turn a dial.

synchronized, so at around two o'clock, bells would have been audible from several directions, one after another, near and far, until all had passed the hour. In the middle of a performance, the three o'clock bells would have been audible too, above the voices of the cast. In Act 2 of *Macbeth*, there is a portentous moment when a bell rings, and Macbeth announces that it is

> *a knell*
> *That summons thee [Duncan] to heaven or to hell.*

No doubt this would have sometimes coincided with the sound of a church bell, which would have added to the dramatic effect.

As for the duration of the plays, there's a clue in the prologue of *Romeo and Juliet* in which the Chorus give a quick synopsis of the plot (spoiler alert!) saying:

The fearful passage of their death=marked love
Is now the two=hours traffic of our stage.

Two hours for *Romeo and Juliet*? The actors would have had to go at a fair lick to get through the full version that quickly, even without an interval. Perhaps the play was shorter when that prologue was written, or perhaps people were so used to time measurement being rather imprecise that they weren't really fussed if the show lasted two hours or three, so long as they were enjoying themselves.

As the afternoon passed, what if an audience member got a little restless and wondered how long this play would be going on for? Most of the public would have to use the sun's position as their guide. But, for the very wealthy, a relatively new invention was available: the pocket watch. First invented in Nuremberg in 1510, these devices relied on internal hairsprings. They had two functions: one was to tell the time, the other was as a show of status. It was the second of these that was more important because, in Shakespeare's time, watches were so unreliable as to be almost useless. They had only an hour hand, no minute hand, and might easily lose or gain two or three hours a day. A minute hand would be pointless. Even those few public clocks that had a dial would show only the hours – London's first public clock with two hands did not appear until the 1670s. Shakespeare was rich enough to own a watch, but he never mentions one.

Nobody measured time in seconds. Life just wasn't that precise. And that's why Shakespeare never mentions seconds either. It was almost as if a minute was as short a time as could be properly grasped. In *King John*, Cardinal Pandulph observes that, having captured Prince Arthur, the king won't have any rest for 'an hour, one minute, nay one quiet breath'. If the public had been familiar with seconds, Shakespeare would surely have added this in for dramatic effect, but instead he goes straight from a minute to an instantaneous breath.

NAVIGATION AND MAPS

Give me the map there. Know we have divided in three our kingdom.

KING LEAR

In the late 1500s, the measurement of tiny quantities was still beyond the reach of everyday technology, and things were equally challenging at the other end of the scale.

Shakespeare grew up in one of the most exciting eras of exploration. New exotic lands were being explored far across the ocean. When he was in his teens, he would have heard of the exploits of Francis Drake, who had just become the first Englishman to circumnavigate the world. A few years later, Shakespeare would have encountered the growing craze of smoking tobacco, which had been brought to London from Walter Raleigh's ill-fated colony in Virginia, in the New World.

Queen Elizabeth's drive to grow trade and explore the world for valuable resources meant more and more travel over greater distances, and that meant a huge growth in seafaring, and with it the need for maps and charts.

Navigation was highly dependent on mathematics. Unless you could accurately calculate your position, there would be a huge inconvenience of delays caused by heading in the wrong direction. Worse than that, you were at risk of being wrecked on rocky shores, or of landing in the wrong place and getting a hostile reception.

There were two fundamental things that a sailor needed in order to navigate: the direction they were pointing in and their location. The first of these could be dealt with by a compass, a device that Shakespeare knew of (they had been around for several hundred years) but only mentions once, in *Coriolanus:*

> . . . *they would fly east, west, north, south,*
> *and their consent of one direct way should be at*
> *once to all the points o' the compass.*

A magnetic needle would point to north, and all you needed to do was read off your bearing from the dial.

Working out your location involved more difficult measurement and calculation. First there was latitude, how far north or south you were of the equator, which could be determined by looking at the sun at noon or the pole star at night and measuring their elevation in the sky. To do this, a mariner would use an angle-measuring device such as a quadrant or an astrolabe, and then refer to an almanac that indicated the elevation of stars on every day of the year. These almanacs would be produced by mathematicians and purchased for use onboard ship.

Longitude – your distance east to west – was much more of a challenge. The crudest method of working it out was simply to measure the distance you had travelled over the water and then plot that on a chart. If you were making steady progress on a calm sea, then you could use a system that was known as 'dead-reckoning'. By letting a rope with regularly spaced knots drop off the rear of the ship, and counting the number of knots that went overboard in 30 seconds, you could work out your speed in (nautical) miles per hour. Navigators worked out that if the gap between knots was just over 47 feet, then one knot per 30 seconds was equivalent to one nautical mile per hour. Twenty miles per hour became known as twenty knots. You could work out your distance by multiplying your average speed by the amount of time you'd been sailing, measured with an hourglass. Clever stuff – but it only worked in gentle conditions. A storm, or strong currents, would mess up your readings and you could find yourself miles from where you thought you were.

A far more reliable guide to your longitude was to look at the position of the stars, which moved across the sky in a regular arc each day. So long as you knew what time it was, the angles to the stars would tell you how far west you were. Unfortunately that's where the problems began.

There were no clocks accurate enough to enable you to work out longitude with any confidence. Even clocks that worked reasonably well on land could not cope with the heaving and rocking of a ship on a swelling ocean. Once land was out of sight, even the most skilled measurements could not guarantee that you would arrive anywhere close to your destination. It would be another hundred years before John Harrison invented a timepiece that was accurate enough to operate reliably at sea.

Shakespeare knew some of the technicalities of navigation. In Sonnet 116, he compares love to the North Star. That's the star which doesn't move in the sky; it was used by sailors to show them the direction of north. Shakespeare refers to it as

an ever-fixed mark . . . whose worth's unknown,
although his height be taken.

Taking the star's 'height' is a poetic reference to working out a position using a cross-staff, which was a simple device used to measure the angle of stars relative to the horizon (it was eventually superseded by the sextant).

A nautical measurement that he often refers to is the fathom, which is used to measure depth. The most famous example is in *The Tempest*, where Ariel tells Ferdinand that

Full fathom five thy father lies.

A fathom is 6 feet, so his father is 5 × 6 = 30 feet under water. 'Full fathom five' sounds much more dramatic than 'thirty feet'. But fathom is one of very few nautical terms in Shakespeare's work.

He makes no mention at all of angles, quadrants, cross-staffs, astrolabes, knots, port or starboard. The relatively small number of references to the technicalities of sailing suggests that Shakespeare might never have sailed on a ship, though he would have been on ferries and wherries across the Thames countless times.

This is confirmed by some of the geographical errors in his plays. In *Two Gentlemen of Verona*, Panthino orders Lance to board the ship that's setting sail from Verona to Milan, which is quite an achievement

Cross-staff in use, from Regimento de Navgacion
by Pedro de Medina, 1552.

given that both cities are landlocked. In *The Winter's Tale*, Bohemia, the Western part of modern-day Czechia, is given a coastline.

There was, however, one nautical innovation that Shakespeare was very aware of.

It had long been known that the earth was a sphere. For practical reasons, however, maps were flat. For maps of a small territory, the curvature of the earth isn't really a problem, but try taking a sphere and flattening it out into a sheet, and you will find that you have to stretch the land near to the poles. The result is a highly distorted map, in which, say, a bearing towards north-west is not represented accurately by an angle of 45 degrees on the map. The mathematical challenge of how to represent the world as a flat map was finally cracked by the Flemish mathematician and cartographer Gerardus Mercator in 1569. His new map allowed navigators to take a bearing and transfer that directly to the ship's compass, so they knew in which direction to sail, and gave his fellow countrymen a huge advantage in their voyages to seek out the riches of the Spice Islands in the East Indies (now known as the Philippines and Indonesia).

Mercator's map wasn't perfect, however, and it wasn't until 1599 that an English mathematician, Edward Wright, corrected Mercator's errors and produced a map upon which sailors could truly rely. The map was crisscrossed with straight lines to indicate points that were on the same bearing (so-called 'rhumb lines'), and this new map was so famous that Shakespeare felt able to mention it in *Twelfth Night*, which was written in 1602. When Maria is joking about the gullible Malvolio, she says

> He does smile his face into more lines than is in the new map with the augmentation of the Indies.

In other words Malvolio's face had even more lines on it than Wright's new map.

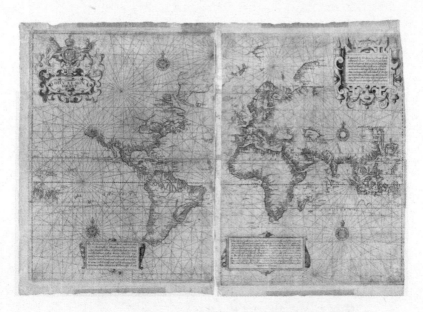

*Edward Wright's 1599 map crisscrossed with faint lines
to show points that were on the same bearing.*

There was of course a more accurate way to represent the position
of distant lands, and that was to present them on a globe. Our friend
Mercator was an expert in making those too, and a Mercator globe
was a prized object in the homes of the wealthy. Who knows, maybe
Shakespeare had one in his home in Stratford. Then again, he was
already part-owner of the biggest Globe of all: his theatre. 'All the
world' was indeed a stage.

CHAPTER VI
MUSIC, RHYTHM AND DANCE

THE MEASURE OF DANCE

Lady: Madam, we'll dance
Queen: My legs can keep no measure in delight

RICHARD II

When Shakespeare used the words 'measure' and 'number', he wasn't always using them in the modern mathematical sense. Often he was referring to rhythm, music and the beats of poetry. In other words, for Shakespeare, mathematics was as fundamental to the arts as it was to the sciences.

Let's start with dance, which was an important feature of plays at the Globe theatre.

In *As You Like It*, when the character Touchstone is explaining that he used to be a courtier, he says 'I have trod a measure.' By this he means that he has learned one of the formal rhythmic stepping dances (a 'measure') that were so popular in court.

The most popular dance mentioned in Shakespeare's plays was the jig, a solo dance with fast footwork. There is also a reference in *Hamlet* to what was later known as a stage jig. At the end of every play at the Globe, the actors would put on a jig as the finale. Even after a tragedy such as *Macbeth*, the musicians would strike up and the actors would assemble for what was often a jolly and sometimes slightly bawdy jig, with the audience clapping and cheering.

Several other Elizabethan dances are mentioned in Shakespeare's plays including the volta, the canario and the pavane. One of the most popular dances was known as the galliard. This dance originated in Italy, but was popularized by Elizabeth I. The underlying rhythm had six beats, a combination of steps and hops, often done alternately by the man and the woman. Watching a galliard is like seeing a couple going through a very choreographed courtship ritual – indeed the romantic use of the word 'courting' originated in the late 1500s, and formal dancing in court might have been partly responsible. In *Twelfth Night*, Sir Toby Belch asks Sir Andrew Aguecheek

What is thy excellence in a galliard?

because that's the sort of skill that will help him in his wooing of Lady Olivia.

Steps and hop in a galliard dance, from Thoinot Arbeau's 1588 book of dance instructions, Orchesographie.

One form of galliard was also known as the cinquepace (pronounced 'sink pace'), from the French word for five and meaning 'five steps'. To dance it, you did four steps and a hop, making five steps in total. To get a feel for the rhythm of a cinquepace, first clap your hands in a six beat like this:

One Two Three Four Five Six, **One** Two Three Four Five Six...
and so on.

Now repeat it, but this time miss out the clap on beat five, so that you get:

One Two Three Four [...] Six, **One** Two Three Four [...] Six...

The gap is the moment when the dancer hops in the air, landing again on beat six.

Later in Sir Toby and Sir Andrew's exchange about dancing, Sir Toby jokes that if Sir Andrew really is good at dancing, he should be doing it all the time. Sir Toby reckons he would even dance when going for a pee in a gutter, or as he puts it: 'I would not so much as make water but in a sink-a-pace' (an Elizabethan 'sink' was a cesspool, the word here being used as a pun on cinque).

This combination of numbers with movement has always been a feature of dance, and most dances are based around twos and threes, as these are the simple numbers that the brain recognizes without any mathematical training. As Feste says in *Twelfth Night* (one of Shakespeare's most musical plays), three is a good number for dancing to:

The triplex, sir, is a good tripping measure.

As we'll see next, simple numbers were the essence of Elizabethan music and, more fundamentally, they were at the heart of Shakespeare's own verse.

PATTERNS OF VERSE

I am ill at these numbers.

HAMLET

When Hamlet declared himself to be 'ill at these numbers' he wasn't boasting about how bad he was at math. Instead he was bemoaning his poor efforts at writing love poetry. It's curious that Shakespeare often uses the word 'number' to mean a verse of poetry,[24] but it shows that he regarded the numbering of lines, and beats within a line, as a fundamental part of his art. Anyone could bash out a bit of doggerel, but writing good verse was recognized as a fine skill. It required something of a mathematical mind, and Shakespeare was one of the finest practitioners of his time.

The numbers five and ten were particularly important to Shakespeare's craft. What is the connection between the following famous Shakespearean lines?

So foul and fair a day I have not seen . . .	(*Macbeth*)
But soft, what light from	
* yonder window breaks?*	(*Romeo and Juliet*)
If music be the food of love, play on . . .	(*Twelfth Night*)
Cry God for Harry, England and Saint George!	(*Henry V*)
A horse, a horse, my kingdom for a horse!	(*Richard III*)

The answer is that they all have ten syllables and share the same rhythm when spoken. Replaced with dees and dums they all sound like this:

Dee-dum dee-dum dee-dum dee-dum dee-dum.

There are five 'dees' and five 'dums', with the emphasis on the dums.

---

24 'Number' is these days used to refer to an individual song or tune in a playlist.

A two-syllable word in which the second syllable is stronger is known as an 'iamb', and this distinctive rhythmic pattern of five iambs is known as iambic pentameter, 'penta' meaning five and 'meter' a measurement. In fact an iamb is an example of what is known in literary circles as a 'foot', so, in stark contrast to what my tape measure says, this is a case where a meter is five feet.

A foot in which the first syllable is emphasized, dum-dee, is known as a 'trochee' (for example, the words 'happy', 'pudding' and 'yellow').

Well over a third of the lines in Shakespeare follow the iambic pentameter rhythm. Why did he choose it ahead of all the other forms that were available? The general view is that English words fit more naturally into iambic form than, say, the trochee form.

It's a bit harder to come up with a natural sounding trochaic pentameter sentence, that is, one where the emphasis is on the first syllable of each foot. Two examples are: 'Would you like to see a play in London?' and 'Hampton Court is very good for tennis'.

When several lines of trochee are strung together the result starts to sound almost hypnotic. Shakespeare uses trochees for chanting, most famously in the spells of the witches in *Macbeth*:

Double, double, toil and trouble
Fire burn and cauldron bubble.

Shakespeare would have learned about different forms of poetic meter from his classical studies: the Greeks and Romans were experts, but one form they didn't use was iambic pentameter. Pentameter was a later invention, popularized by the great English writer Geoffrey Chaucer in the 1390s in his classic *Canterbury Tales*. At times Chaucer's verse is almost chatty, with lines such as this from the Prologue: 'Some twenty years of age he was, I guess.'

Although iambic pentameter is the famous rhythm within Shakespeare, the majority of his lines do not use this meter at all.

Generally he reserves pentameter for his rich and noble characters, while the poor and common folk speak in plain text. This is the word equivalent of what we saw him do with numbers, where he saved precise numbers for the nobility while the commoners got rounded numbers. Of course the exception proves the rule. Perhaps the most famous Shakespearean quotation is the famous Hamlet speech:

To be or not to be that is the question.

There are five 'dee-dums' and then, tacked onto the end, an eleventh syllable (an extra 'dee'). This is known as a weak ending, and used sparingly it was deemed to be an acceptable variant of iambic pentameter, a bit like half-rhymes in a rhyming poem.

Interestingly, in the first published quarto of *Hamlet* (in 1603), the opening line of the Hamlet speech was different:

To be or not to be ~ ay there's the point.

That's ten syllables in a perfect iambic pentameter! So why did Shakespeare change it? It's possible that this early version was not actually Shakespeare's version at all. Sometimes bootleg versions of his plays were published, and the printed words were the half-recollections of actors or an audience member rather than Shakespeare's original. Whatever the explanation, most people reckon the later version is better, despite its slight rhythmic impurity.

RHYMING PATTERNS

I thank you for this profit, and from hence
I'll love no friend, since love breeds such offence.

OTHELLO

The other pattern that is embedded within Shakespeare is rhyme – though when it comes to his plays, relatively few lines actually rhyme with each other. It's sometimes said that Shakespeare uses rhymes when a character is either leaving or loving.

For example, here's Richard III at the end of the first scene, summing up where we've got to in the plot so far:

Clarence still breathes, Edward still lives and reigns
When they are gone, then must I count my gains. (Exits)

'Leaving' also includes departing this life. Here's Buckingham later in the same play, about to be executed:

Come sirs convey me to the block of shame
Wrong hath but wrong and blame the due of blame.

And, as for love, look no further than *Romeo and Juliet*, where there is love and rhyme aplenty:

Did my heart love till now? Forswear it, sight.
For I ne'er saw true beauty till this night.

However, the more interesting and sophisticated rhyming patterns are to be found in Shakespeare's narrative poems and his sonnets. His first published work was the narrative poem *Venus and Adonis*, which begins:

Even as the sun with purple-colour'd face
Had ta'en his last leave of the weeping morn

Rose-cheek'd Adonis hied him to the chase
Hunting he loved but love he laugh'd to scorn
Sick-thoughted Venus makes amain unto him
And like a bold-fac'd suitor gins to woo him.

Not only is almost every line written in iambic pentameter, but every one of the 199 verses has six lines in the rhyming pattern of ABABCC.

His next poem was *The Rape of Lucrece,* a 265-verse epic, again in iambic pentameter, but this time each verse was seven lines in the rhyming pattern ABABBCC.

Finally, there were Shakespeare's 154 sonnets, each exactly fourteen lines long, with the same iambic rhythm and rhymes in the format ABABCDCDEFEFGG – three clusters of four, with a rhyming couplet to finish.

There was no rule that said a poet had to follow patterns like this, but different rhyming patterns create different moods, and the constraint of fitting into a format can make the outcome more satisfying for the reader and the writer – much as the best crosswords fit into a symmetrical grid, not because they have to, but because they are more aesthetically pleasing.

The numbering of the sonnets is also worth a mention, as some of the numbers fit the poem's theme. Meanwhile, Sonnet number 12 begins:

When I do count the clock that tells the time.

Twelve is the number of hours on a clock, which was surely deliberate. Meanwhile, Sonnet 8 begins

Music to hear, why hear'st thou music sadly?

Why eight? That's the number of notes in an octave. What we don't know is whether it was Shakespeare or his printer who decided on the order of the sonnets and made the numbering significant. But mention of octaves takes us seamlessly on to music, the most mathematical of all the arts.

MUSIC AS A
MATHEMATICAL SUBJECT

A *tutor . . . cunning in music and the mathematics*

THE TAMING OF THE SHREW

This quotation is a clue to where music belonged in sixteenth-century culture. Why has Hortensio disguised himself as a mathematics and music tutor? Is this because he is a jack of all trades? Not at all. Remember, music and mathematics were one and the same thing, as music was one of the four mathematical subjects in the Quadrivium.

Music was extremely important to Shakespeare. The University of Birmingham hosts a database of all the musical references in Shakespeare's plays and poems. There are over two thousand in total, and around a hundred songs. Add to that the musical jigs at the end of plays, and it's clear that Shakespeare must have loved music and been very musical himself.

The idea of music as a mathematical science dates back to Pythagoras in ancient Greece. The story goes that Pythagoras was walking past a foundry where an ironmonger was beating two metal pipes. The clang coming from the first pipe made a note, and the clang from the second pipe made a note that sounded the same – but higher. Curious to know more, Pythagoras noticed that the first pipe was twice the length of the second, the two notes making a pleasing harmony that we would now call an octave. Pythagoras was one of many ancient philosophers who believed that the universe was built on simple numbers, and when he found that pipes in a ratio of 3:2 also made pleasing harmonies, he set out to create a set of musical notes that were all related to each other in the ratio of 2:1, 3:2 and 4:3. He didn't quite manage it, but he did find that the octave could be spanned in seven steps, creating intermediate notes that were fairly evenly spaced,[25] in what we now call a scale.

The easiest way to make notes was by plucking a string. By dividing the string into different fractions you got octaves and other notes in between (what we'd now call perfect fifths, major thirds and the rest), just as a modern guitarist or violinist does today.

Is this the 'sweet division' that Juliet is referring to when she curses the lark for bringing in the dawn, meaning that Romeo has to leave her?

It is the lark that sings so out of tune,
Straining harsh discords and unpleasing sharps.
Some say the lark makes sweet division
This doth not so, for she divideth us.

Possibly. Or perhaps it is a double musical pun, because in Shakespeare's time there was a new fashion for adding jazzy extra notes into tunes, which was known as 'division' (literally dividing a beat up into several beats).

Musical instruments in Shakespeare's time were still tuned using the simple Pythagorean ratios mentioned earlier. On the stage, in court and in the streets there were instruments that could be tuned, such as lutes and viols, with strings that could be tightened to change the tuning. Other instruments such as shawms (an early form of oboe), pipes and cornets had holes drilled in fixed positions, and tuning could not be adjusted. Occasionally these instruments would sound awful together.

Even with just a single instrument, not every pair of notes sounded harmonic, and Shakespeare often refers to clashing notes and disharmony. In *Richard II* the king talks of a musician who 'knows no

---

25 The classic octave actually has five steps of roughly the same size (known as tones) and two half-steps (semitones).

touch to tune the harmony'. Music education was partly the instruction in how to use ratios that would produce 'nice' tunes.

Tunes were often written to deliberately avoid clashing notes; meanwhile, musicians and university mathematicians were busy researching the best combination of ratios that would produce harmonious tunes, whichever key you started on. It was a complicated and messy business, and I will avoid going down the rabbit hole of the history of music theory.

In simplistic terms, if you study music today, you will learn about major and minor keys, but in the 1500s, music was played using a smaller range of notes, and played in what were called modes. In effect, modal music consisted of tunes that you can play using only the white keys on a modern piano. This included a number of tunes that are still familiar today, such as 'Scarborough Fair' (made famous by Simon and Garfunkel in the 1960s) and 'Greensleeves'. The latter is mentioned twice by Shakespeare, including a slightly nonsense line from Falstaff in *The Merry Wives of Windsor*:

Let the sky rain potatoes; let it thunder to the tune of Green Sleeves.

There were seven different modes, depending on the note in the scale that you started on. By far the most popular mode in Shakespeare's time was what was known as Dorian mode,[26] which (coincidentally) starts on the note that we now call D. If you have access to a keyboard, you'll be able to confirm that you get the tune of Greensleeves if you play the white keys D F G A B A G E C. If, however, you start on any other note, you won't be able to reproduce the tune unless you cheat and use at least one black key.

---

26 There is more information on the Dorian mode in the appendix, page 188.

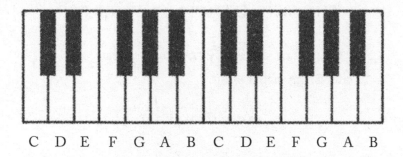

Section from a modern keyboard.

Another tune from Shakespeare's time that remains popular to this day is 'Three Blind Mice'. It was written (or at least published) by a young Cambridge music graduate by the name of Thomas Ravenscroft in 1609, a couple of years before Shakespeare wrote *The Tempest*. 'Three Blind Mice' is a round in which four singers each sing the same melody beginning at different times. Shakespeare called this type of song a 'catch'. Although there are catches in *The Tempest*, there is no mention of 'Three Blind Mice', but it was a popular song and it's almost certain Shakespeare would have heard it, perhaps being sung in the street or a pub. But how did it sound?

On a modern keyboard, 'Three Blind Mice' is based around the note C, and begins E D C. But if Shakespearean music was usually written in Dorian mode, 'Three Blind Mice' should be based around D, not C, and begin F E D instead. Try playing that on a keyboard and you get a rather downbeat, sad tune, very different from the jolly modern nursery version. And yet, remarkably, that sad version is exactly how the original sounded. Have a listen online to 'Three Blind Mice – Ravenscroft' and you'll hear a doleful song about the miller's merry old wife scraping the tripe and getting you to lick the knife.

THE MUSIC OF THE SPHERES

And certain stars shot madly from their spheres,
To hear the sea-maid's music.

A MIDSUMMER NIGHT'S DREAM

M usic was regarded as perhaps the purest and most sublime of art forms, and since the medieval view was that the heavens were created by God, heaven too must have music. In *Henry VIII*, Queen Katharine refers to it as 'celestial harmony'.

The idea that music was linked to the heavens went back to the time of Plato. It was believed that the same rules that governed sweet music on Earth must apply to the stars, but that we humans with our mortal imperfections could not hear it.

In fact, Pythagoras and his followers differentiated three types of music: the music of instruments, the music of the human body and soul, and the music of the cosmos. His reasoning was that, because objects produced sound when in motion, planets moving in orbit should also produce a sound. Why was this sound inaudible? Pythagoras and his followers believed that, because this sound was constant, humans had nothing to compare it with and therefore could not distinguish it from our known idea of silence.

This is what Lorenzo says in *The Merchant of Venice* as they gaze at the stars (orbs):

There's not the smallest orb which thou behold'st
But in his motion like an angel sings,
Still choiring to the young-eyed cherubins;
Such harmony is in immortal souls;

But whilst this muddy vesture of decay
Doth grossly close it in, we cannot hear it.

This heavenly music had a mathematical name. It was known as the 'music of the spheres'. The spheres were the celestial frame in which the stars and the planets were held. In Shakespeare's world, mathematics, music and astronomy were inextricably linked.

CHAPTER VII
ASTRONOMY AND ASTROLOGY

TELESCOPES AND HOROSCOPES

Not from the stars do I my judgement pluck
And yet methinks I have astronomy

SONNET 14

Planets and stars featured far more prominently for the people in Shakespeare's world than they do for most of us today. For a start they were more visible. There was no light pollution, so on cloudless nights the sky would have been filled with visible stars. The stars had a practical use too, as they provided a sense of direction for anyone walking at night. The patterns formed by the stars would have been familiar to everyone – from daily exposure and from the mythical tales of Roman Gods such as Jupiter and Venus. And most people would have known of the link between different constellations in the sky to the signs of the Zodiac – Gemini, Libra and so on – even if not everyone knew exactly where in the sky these constellations were.

Today astronomy and astrology are treated as completely different things. Astronomy is the science of stargazing. Astrology is the mystical practice of predicting everyday events and behaviours based on the movements of the planets and the stars. For one you use a telescope, for the other a horoscope. However, in Shakespeare's day, the two words were interchangeable (though he never used the word 'astrology' himself).

When in *Troilus and Cressida* we hear that 'astronomers foretell it', it's because it was generally believed that what happened in the heavens would enable astronomers to predict what would happen on Earth.

So intertwined were the two disciplines that some of the most rigorous and talented mathematicians of Shakespeare's time – such as John Dee, of whom we'll hear more later – were making detailed astronomical measurements and calculations, and then using them to advise the new queen on which date she should hold her coronation, and where to send an expedition to the Americas.

We can tell how interested Shakespeare was in this field from the number of times he refers to the stars, planets, sun and moon in his work. The table below compares Shakespeare's count with that of several other leading playwrights and poets of his era. His score of 510 mentions is almost five times that of John Milton, who comes second. It's not entirely fair as a comparison because Shakespeare has more surviving works than some of his rivals, and his career was longer too. But still, it seems that, when it came to looking at the heavens, Shakespeare was the frontrunner.

Playwright	Lifespan	No. of mentions of stars/planets/moon in their work
William Shakespeare	1564–1616	510
Christopher Marlowe	1564–1593	57
Ben Jonson	1572–1637	42
John Donne	1572–1631	55
John Milton	1608–1674	118

From *Shakespeare's Astronomy*, Michael Rowan-Robinson.

Let's take a step back and consider how most people in sixteenth-century Europe believed that the universe was structured. Fundamental to it all was the idea that the earth was at the centre of everything, and that the heavens rotated around us. This was a view that had been established by the ancient Greek mathematician Ptolemy, and completely embraced by the Church. Any other theory was a heresy that could lead to imprisonment or death.

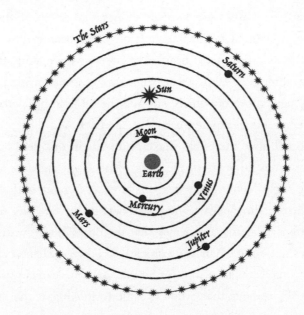

According to Ptolemy, the earth sat at the centre
of eight orbiting celestial spheres.

As Ptolemy saw it, the universe comprised eight transparent spheres with the earth at their centre. Each sphere carried a different celestial body that revolved around the earth in 24 hours. The first seven spheres carried the 'planets' (more of which anon), and beyond them were the fixed stars. It was deemed impossible that anything could move between these spheres, so when there was any unusual disturbance in the night sky – a comet or a meteor, for example – this was a huge and portentous event that must surely carry deep meaning. In *A Midsummer Night's Dream,* when Oberon remembers hearing the beautiful song of a mermaid, he says

And certain stars shot madly from their spheres.

to indicate just how significant a moment this was.

Altogether there are twenty-two mentions of spheres in Shakespeare's work. Added to fourteen mentions of globes, this is comfortably the most frequently mentioned mathematical shape.[27]

At Hampton Court there is a clock built in 1540 that beautifully illustrates the world of Tudor astronomy.

The clock at Hampton Court, built in 1540.

---

27 In contrast, Shakespeare never mentioned triangles. To be fair, few if any other playwrights did either, despite the fact that love triangles were a key element in many romantic plots.

Like most clock faces of its time, there is only a single hand, to mark the hour. Unusually it is a 24-hour clock, so it maps out the whole day, but it's the decorations that are particularly interesting. In the outer ring are the signs of the zodiac. At the centre of the face is a circle representing earth. Just above and to the right of the earth is a disc representing the moon, which changes each day to indicate the moon's phase (when this photo was taken, there was a crescent moon). Most fascinating of all is the hour hand itself. On it is a symbol of the sun, which dutifully completes its orbit of the earth over 24 hours every day.

So important was astronomy to Shakespeare's company of actors that the ceiling of the Globe theatre was painted with a tableau representing all the heavenly bodies. When Hamlet says:

this most excellent canopy, the air, look you this brave o'erhanging firmament, this majestical roof fretted with golden fire,

he is referring both to the real sky and to the painted heavens above the stage.

The stars and the planets weren't simply there to be looked at in awe. It was widely believed that the mood and health of a person were directly affected by 'rays' being emitted by these heavenly objects. The word 'influenza' comes from the belief that the alignment of the stars and planets could actually 'influence' your health.

RULED BY THE SEVEN PLANETS

The wars have so kept you under that you must needs be born under Mars.

ALL'S WELL THAT ENDS WELL

The idea that there were seven planets was at least 2000 years old by the time Shakespeare first encountered it, and there had been almost nothing to suggest this was anything but the absolute, unchallengeable truth.

The word 'planet' comes from the Greek word *planetes* meaning a wanderer. The planets were those bodies in the sky that moved differently from the fixed pattern of stars. Early astronomers had noticed that not only did these bodies move at a different rate, but they sometimes appeared to loop back on themselves.

If you try to guess what those seven planets were, you might get a surprise. Yes there were the usual suspects: Mars, Venus, Mercury, Saturn and Jupiter. But the other two? They were the moon and the sun, or as Shakespeare calls it in *Troilus and Cressida*,

The glorious planet Sol, In noble eminence enthroned and sphered.

Neptune and Uranus were too small to be visible to the naked eye.

The fact that there happened to be seven of these so-called planets was a lucky fluke, because ever since people began to count, they had noticed that the number 7 was different from the other numbers around it. Our counting system is based on 10 (because that's how many fingers we have), and all the other numbers in the range 1 to 10 have some connection to each other: 4, 6, 8 and 10 are all multiples of 2, 10 is twice 5, 6 and 9 are multiples of 3, and 1 divides into everything.

That just leaves 7, which is out on its own, and that made it special. The fact that there were, it seemed, exactly seven wandering planets would have struck early astronomers as more than just a coincidence.

The planets became associated with Gods, and also, from Babylonian times, with the days of the week. Between the English and French languages, our week is still named after those seven ancient planets:

Moon Day	Monday
Mars Day	Mardi (Tuesday)
Mercury Day	Mercredi (Wednesday)
Jupiter Day	Jeudi (Thursday)
Venus Day	Vendredi (Friday)
Saturn Day	Saturday
Sun Day	Sunday

The seven planets had (according to mystical beliefs) another connection with earthly life. One of Shakespeare's most famous uses of the number 7 appears in *As You Like It*, in the 'all the world's a stage' speech, where Jaques runs through the seven ages of man. Although Shakespeare doesn't mention it, each of these ages had historically been associated with one of the planets. Shakespeare's ages don't follow the planetary path precisely, but they have much in common, as shown in the table on page 115.

Why were the ages linked to planets in this order? It was all based on the time it took for each planet to return to the same position relative to the stars (the planetary 'year'). The idea was that life slows down with age. The shortest 'year', linked with infancy, was the moon with its 24-hour cycle, and the longest was Saturn at nearly 30 years. The sun's year was of course 365 days, the same as our Earth year.

Age	Planet[a]	Shakespeare's Seven Ages of Man
Infancy	Moon	At first the infant, Mewling and puking in the nurse's arms
Childhood	Mercury	Then the whining school-boy, with his satchel And shining morning face,
Adolescence	Venus	And then the lover, Sighing like furnace, with a woeful ballad Made to his mistress' eyebrow.
Young adulthood	The sun[b]	Then a soldier,
Maturity/ the soldier	Mars[c]	Full of strange oaths, and bearded like the pard,
Justice and wisdom	Jupiter	And then the justice, In fair round belly with good capon lin'd
		The sixth age shifts Into the lean and slipper'd pantaloon, With spectacles on nose and pouch on side
Old age	Saturn	Last scene of all, That ends this strange eventful history, Is second childishness and mere oblivion

a The number 7's grip on the physical world didn't stop with the planets. At the time, only seven metals were known: gold, silver, copper, iron, mercury, tin and lead. Each metal was associated with a planet – gold was the sun, silver the moon (we still call it 'the silvery moon'), Venus copper, Mars iron, Mercury itself, Jupiter tin and Saturn lead.

b The sun typically represented the fourth, peak phase of life, when a man was at full energy, followed by Mars. Jupiter would usually be the sixth, 'wisdom' phase but is possibly covering justice and wisdom here. It's all poetic licence on Shakespeare's part.

c Mars is the origin of the military word 'martial'.

Shakespeare makes copious mentions of the planets, or the Gods associated with them. Despite the fact that many of the plays have underlying love stories, Venus is actually the least quoted of the planets (a mere 30 mentions), some way behind Mars (45 or so), so when it comes to Shakespeare, war beats love. Both, however, are a long way behind the moon and the sun.

There are numerous references to the influence of the seven planets throughout the plays, for example when Henry V asks

Saturn and Venus this year in conjunction ~ what says almanac to that?

But it was stars rather than planets that Shakespeare seemed to call on most often when referring to fortune, with mentions of 'luckiest stars' and 'good stars'.

Twice Shakespeare refers to the 'seven stars'. These are the Pleiades, a cluster of seven stars that were sometimes associated with death and mourning. Once again, the fact that there were seven of these stars gave them extra astrological significance.

But did Shakespeare himself believe there was a deep mystical meaning to the number 7? There's a quote in *King Lear* that suggests he was sceptical. In one scene, the Fool is posing jokey riddles to the king. At one point he says

The reason why the seven stars are no more than seven is a pretty reason.

He's expecting King Lear to reply 'OK, tell me, what is the reason?' but, for once, the king comes back with a witty riposte:

Because they are not eight?

This deadpan answer is funny, because it suggests that it's silly to read any meaning into the number of stars; it's just a coincidence that there happen to be seven of them.

And, later in the same play, there's an even stronger hint that Shakespeare was a sceptic about astrology. At one point Edmund says it's time to stop blaming the stars for our misfortunes, and instead take responsibility for our own behaviour.

> When we are sick in fortune, often the surfeit of our own behaviour,
> we make guilty of our disasters the sun, the moon, and the stars; as if
> we were villains on necessity; fools by heavenly compulsion;
> knaves, thieves, and treachers by spherical pre=dominance;
> drunkards, liars, and adulterers by an enforc'd obedience of
> planetary influence; and all that we are evil in, by a divine
> thrusting on. An admirable evasion of whore=master man, to lay
> his goatish disposition to the charge of a star!

King Lear was written around 1606, towards the end of Shakespeare's career. Maybe he was becoming a bit cynical about the more outlandish predictions that were attributed to the movement of the stars and planets. Or perhaps he was learning about the new discoveries that were being made by astronomers across Europe, which were turning the old-world certainties of the universe upside down. Word was spreading: whisper it gently, but perhaps the earth was not at the centre of the universe after all.

THE SCIENCE OF ASTRONOMY

O, learn'd indeed were that astronomer
That knew the stars as I his characters

CYMBELINE

As we saw in Chapter 2, astronomy was one of the seven Liberal Arts (that number again!), and one of the four mathematical subjects in the Quadrivium. The purpose of this mathematics was to measure the positions of the stars and to plot their future trajectories, both for practical reasons (such as navigation) and for astrological ones (to tell somebody's fortune). Both these purposes called for a level of precision that was above that of any other walk of life. For this reason, astronomy was one of the two practical applications of math that were driving the advance of the subject. (The other was gambling, which we met in Chapter 3.)

The model of the universe produced by Ptolemy had served well, but it struggled to explain the loopy motions followed by some of the planets. In the first part of *Henry VI* (written around 1592) Shakespeare reveals that nobody quite understood what was going on.

Mars his true moving, even as in the heavens,
So in the earth, to this day is not known.

This was, however, all about to change. One of the great coincidences in Shakespeare's life is that he was born in the same year, 1564, as Galileo Galilei, one of the most influential astronomers of all time. Galileo was born in Italy. His father was a musician, so the music of the spheres would have been a natural topic of conversation around the dinner table. At university he began studying medicine, but his true interest was in science, mathematics in particular, and that was the foundation of his famous career as an astronomer.

Galileo would build on the discoveries of astronomers who came before him. Earlier in the century, the Polish astronomer Nicolaus Copernicus had become increasingly sceptical about the idea of an earth-centred (or geocentric) universe. His astronomical measurements just didn't support it. Finally, Copernicus came up with a completely different model, in which the earth spun on its axis once a day and orbited the sun once per year (a sun-centred or heliocentric universe). But challenging the geocentric model meant he would also be challenging the Church, which would lead to embarrassment, anger and almost certainly retribution. He therefore waited until he was on his deathbed in 1543 before publishing the results, in a book entitled *de Revolutionibus* or 'On the Revolution of the Celestial Spheres'. The Catholic Church was livid, but mathematicians quietly observed that Copernicus' calculations in his planetary tables seemed to give more accurate predictions of planetary motions than those based on the old model.

Next to pick up the baton was a Danish nobleman by the name of Tycho Brahe, who had two great passions in life: astronomy and parties. The latter would finally lead to his early demise – after a sumptuous banquet in Prague it's believed that Brahe's bladder burst, and he died ten days later. Before that grisly end, however, he had set about making incredibly detailed measurements of the positions of stars and planets in the night sky. At his disposal was a huge quadrant (like a giant school protractor) from which he could measure angles to within 1/60 of a degree (a 'minute'), but all his observations were made with the naked eye.

Working in his remote island observatory, Brahe was largely free from interference from snooping churchmen, and his meticulous scientific measurements also backed up the idea of Copernicus' heliocentric universe (albeit with his own rather quirky alternative in which five planets orbited the sun, but the sun and moon orbited the earth).

Tycho Brahe depicted using a 'mural quadrant'. A star or planet was observed through a slit in the wall and its angle of elevation was then be read from the scale around the quarter circle.

Brahe was fortunate that, while still in his mid-twenties, he was witness to a rare and spectacular event in the skies. It was the appearance in 1572 of what appeared to be a new star, and it was so bright and obvious that everyone, including the young William Shakespeare, would have seen it. The English astronomer Thomas Digges recorded his first sighting in a diary:

> The 18th of November in the morning was seene a star northward verie bright and cleere . . . in bigness it seemed bigger than Jupiter . . . it was found to be far above the moone than ever anie comet hath beene seene.

It was what we would now call a supernova (the huge explosion when a star reaches the end of its life), and everyone in Elizabethan England seems to have been spooked by it. It's the sort of one-off event that would have stuck in Shakespeare's memory throughout his life, and there are some who think that it is mentioned at the start of *Hamlet*. In the opening scene, two soldiers standing guard at Elsinore Castle are talking about a ghost that they have seen. They describe the moment when it happened:

> When yond same star that's westward from the pole
> Had made his course to illume that part of heaven
> Where now it burns.

Was this bright burning star harking back to the supernova? If the opening scene takes place in November (as some academics have speculated), the supernova was in the right place in the sky at the right time of day for it to be the star that was 'westward from the pole'. But nobody can be sure.

There are intriguing clues in *Hamlet* that Shakespeare had heard about Tycho Brahe's work. Expressing his love for Ophelia, Hamlet writes

Doubt that the stars are fire
Doubt that the sun doth move . . .
But never doubt I love.

What does he mean by 'doubt that the sun doth move'? In essence Hamlet is saying, 'Here are some things that you are allowed to doubt, but don't doubt that I love you.' But the stars *are* made of fire, so doubting that would be bordering on heresy. By the same token, are we supposed to think that doubting that the sun moves (i.e. suggesting that instead it is at the centre of the solar system) is also heresy? Maybe. But equally, it could mean that it's obvious that the sun moves through the sky during the daytime (we all know this, even though today we know it's not really the sun that is moving), and that this is the truth that's being doubted. People argue both ways on it.

The final clue is more tantalizing. In 1590, Tycho Brahe wrote to a friend in London, and included with his letter four copies of his portrait that had recently been engraved in Amsterdam. These portraits would have been circulated in astronomical circles around London, and were almost certainly known to Thomas Digges, whose family lived close to Shakespeare in London. Around the portrait are the names of Brahe's illustrious relatives, including two with the names Rosencrantz and Guildenstern, who are two characters in *Hamlet*.[28] Was a sighting of Brahe's portrait the inspiration for Shakespeare when looking for names for his Danish play?

The links with Tycho Brahe remain open to debate, but there is less disagreement about a link between Shakespeare and Galileo.

Galileo's major contribution to astronomy was the invention of a telescope that enabled stargazers to view things that had simply never been seen before. In 1609, while studying Jupiter, Galileo observed four

---

28 Made famous in Tom Stoppard's play *Rosencrantz and Guildenstern Are Dead.*

A 1586 portrait of Tycho Brahe by Jacques de Gheyn. His noble ancestors Rosenkrans and Guldensteren are included in the surrounding shields.

moons orbiting the planet. Here for the first time was proof that there were more than seven 'planets' in the sky, and confirmation that the Greek model of the universe was not fit for purpose. Galileo published this view. Once again, the Catholic Church was apoplectic; Galileo was eventually forced to recant and was put under house arrest to prevent him from publishing any more of his dangerous findings.

But the secret was out. And roughly a year after Galileo's moons had become common knowledge came the first performance of Shakespeare's new play *Cymbeline*. In the play, there is a curious scene in which the God Jupiter descends from the ceiling, while around him four ghosts walk steadily in a circle. There is no explanation of what is going on, but the symbolism seems clear. Four objects orbiting around Jupiter. This can't be a coincidence. If he had had any reason to doubt the new astronomy beforehand, it seems Shakespeare had now bought into the idea.

THE NEW CALENDAR

Thirty dozen moons with borrowed sheen
About the world have times twelve thirties been

HAMLET

S hakespeare's description of a year lasting 'thirty dozen moons' is typical of his creativity with numbers. The moon takes a day to orbit the earth, so Shakespeare is describing a year as 30 × 12 moons = 360 days. He's using a little poetic licence because there are of course 365 days in a year, but Shakespeare didn't have a pithy way of saying 365. Given his love of scores, the only surprise is that he didn't call a year 'eighteen score moons' instead of 'thirty dozen moons', saving himself a syllable in the process.

In fact, the moon takes slightly under 30 days to go through its full cycle – it's more like 29½. Full moon to no moon is half of that, 14¾. That number 14 is significant, because it is two lots of 7 – that mystical number yet again. This is probably why, 3000 years ago, the ancient Babylonians settled on seven days in a week, because a month (literally a moon-th) was very close to four lots of 7. Furthermore, the number of days in a year, 365, was only a day off being a multiple of seven (365 is 52 weeks plus one day). Add to this the fact the number of months, 12, is an extremely handy number when dividing things up (as we saw with money) and our modern calendar was set in stone.

Well ... almost. The problem is that the earth takes a few hours longer than 365 days to orbit the sun. This means that, over time, the date of the winter and summer solstice (the longest and shortest days of the year) will change, and after a few centuries, the calendar will no longer reflect the right seasons. The Julian[29] calendar had introduced

---

29 This calendar was introduced by Julius Caesar in 45 BC.

the idea of an extra day every four years, and had been formally adopted by the Catholic Church in AD 325, but even these leap years were not enough. By the late 1500s, slippage had been going on for over 1200 years, and the error now amounted to ten whole days. The Church was particularly concerned that Easter was now shifting towards summer, and there was a risk that it might start to coincide with other festivals. Something had to be done.

In 1582, Pope Gregory XIII pronounced that the calendar should skip forward ten days, and that everyone should adopt his new Gregorian calendar which, by making a small adjustment to the frequency of leap years, would prevent any future slippage of the calendar. But in England, which was now Protestant, the idea of being told what to do by the Catholic Church was unthinkable. The queen's mathematician-in-chief John Dee was asked to come up with his own proposal. Unfortunately Dee agreed with the Catholic Church's findings, but fortunately for the English establishment, he decided that the Catholics had made a crucial error in setting the year 325 as the starting date of their calendar. Instead, Dee advised that the calendar should be shifted by *eleven* days, as this would mean that it was aligned with the position that the sun would have been in during the year of Jesus' birth.

The queen and her advisers were inclined to agree with this solution, but the Anglican Church was having none of it. Even an eleven-day shift would give the impression that they were bowing to the will of the Catholic Church. So it was decided that England would stick with its existing calendar,[30] while much of continental Europe made the switch. For anyone travelling from London to (say) Antwerp at that time, it wasn't a matter of moving your clock on by an hour, but rather moving your calendar forward by ten days. It must have been a constant source

---

30 It wasn't until 1752, 170 years later, that Britain finally adopted the Gregorian calendar. By now there was an 11-day difference, so the day after Wednesday 2 September 1752 was Thursday 14 September.

of confusion, and possibly even legal complications: if a document had been dated 12 October, you would need to check if this was according to the English or the European calendar.

This calendar mismatch must have been a significant talking point for anyone who travelled abroad in the late 1500s (merchants, scholars, explorers), but Shakespeare makes no obvious reference to it. However, there is a line in the First Folio of Shakespeare's plays that suggests perhaps he did make one joke about the calendar confusion. While plotting the murder of Julius Caesar, Brutus asks his servant:

Is not tomorrow the First of March?

If you know your Shakespeare, you are probably thinking: surely it's the *Ides* [fifteenth] of March, not the first, as in the famous line 'Beware the Ides of March.' Editors of later editions assumed this First Folio line was a mistake, either by Shakespeare or by the printer, and corrected it to have Brutus asking if tomorrow was the Ides of March. But perhaps it wasn't a mistake at all, and Brutus was worrying if he'd got the wrong date. Whether Shakespeare meant it or not, if 'the first of March' was the line delivered by the actors in the original Globe production, the more worldly members of the audience would have understood only too well why Brutus was getting confused with his calendar. It might even have got a laugh.

CHAPTER VIII

COLOURS AND THE RAINBOW

RAINBOWS IN SHAKESPEARE'S WORLD

What tellest thou me of black and blue?
I was beaten myself into all the colours of the rainbow.

THE MERRY WIVES OF WINDSOR

The motion of the stars and planets might have been a source of wonder at night, but daylight sometimes brought its own equally awe-inspiring spectacle in the heavens – the rainbow. Rainbows were generally regarded as a good sign; they were a symbol of new beginnings and also of peace. The Romans called the rainbow Iris, and according to myths she was the messenger between heaven and earth.

As with the solar system, the science and mathematics behind rainbows had been misunderstood since ancient times, but during Shakespeare's lifetime that was all to change.

Shakespeare refers to rainbows several times – sometimes in the form of Iris, the rainbow goddess. However, he never mentions which colours are in a rainbow, or the number of colours. So when Falstaff declares that he has been beaten into all the colours of the rainbow, how many can he see? And which ones?

As children we are taught that there are seven colours in a rainbow: Red, Orange, Yellow, Green, Blue, Indigo and Violet. There are a couple of mnemonics to help remember them, of which my favourite is:

Richard Of York Gave Battle In Vain.

This is a reference either to King Richard III (he of the Shakespeare play) or, more likely, to that king's father, another Richard, who also ended up on the losing side in battle, as described in the final part of *Henry VI*.

However, these seven colours were only formally established in 1704. *The Merry Wives of Windsor* was written almost exactly one hundred years before this. If Shakespeare (through the voice of Falstaff) had been forced to reveal exactly how many colours there were in his rainbow, he would almost certainly not have said seven. In the sixteenth and seventeenth centuries, the colours in the rainbow were more a matter of personal taste and interpretation.

That is not to say that people hadn't tried to count them. The first person to seriously study and try to explain rainbows seems to have been Aristotle in around 350 BC. And, according to Aristotle, there were three colours in a rainbow. Here's what he wrote in his book *Meteorologica*:

There are never more than two rainbows at one time. Each of them is three-coloured; the colours are the same in both and their number is the same, but in the outer rainbow they are fainter and their position is reversed.

The three colours that Aristotle chose to see were red, green and a bluey-purple[31] (ancient Greek *íon*, which today we would call violet, after the flower). These were the only colours that he believed could not be created by mixing other pigments. Aristotle acknowledged that there was also a yellow colour that could often be seen between the red and green, but he thought this was just a trick of the eye.

Why this fixation with three colours? It was based on a philosophical idea of symmetry that Aristotle had picked up from Plato, who believed that two opposites needed a third in the middle to unite them.

Thanks to Aristotle's influence on thinking, the idea of three colours in the rainbow persisted for many centuries and was still popular in Tudor times. With his classical education, Shakespeare himself might have been aware of this nugget of ancient scientific wisdom.

The notion of three colours was reinforced in art too. A painting from 1552, about ten years before Shakespeare was born, entitled *Das Zeichen des Bundes* ('The Sign of the Covenant'), which describes

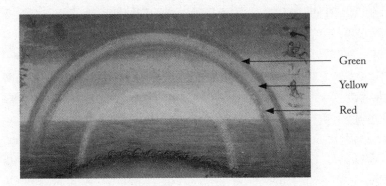

Green

Yellow

Red

'Das Zeichen des Bundes' from the Augsburger Wunderzeichenbuch *(1552), showing a three-colour rainbow.*

---

31 Aristotle was in some ways two millennia ahead of his time, as these are the three colours used to make up the pixels on a computer screen. You also obtain these colours when you combine pairs of the 'CMYK' colours used for printers: C(yan) + M(agenta) = Purple, C + Y(ellow) = Green, M + Y = Red.

the meaning of the rainbow, clearly shows three bands. The original painting was in colour of course, but even in the greyscale version shown here, the three bands in the main rainbow are clearly visible. However, the artist has mixed up the ordering of the colour bands. The top band is green (instead of red), the middle band is yellow and the bottom band is red (instead of green). There's no blue, and the fainter second rainbow is on the inside rather than the outside, where it should be.

There was a more famous rainbow painting that Shakespeare might well have seen in person when he was performing in royal circles. Around 1600, when Shakespeare was in his prime, Isaac Oliver produced the so-called rainbow portrait of Elizabeth I. In it, the queen holds what seems to be a translucent rainbow, next to which is the inscription 'Non Sine Sole Iris' (without the sun there is no rainbow). The rainbow here is representing peace, while the queen is of course the sun. The rainbow in the painting has very little colour, but it appears to have three distinct bands.

Not everyone agreed on the three-coloured rainbow, however. The medieval writer Bartholomaeus Anglicus was one of several to suggest that a rainbow had four colours − in his case red, brown, blue and green. The argument for four was that the colours symbolized the four elements: fire, air, water and earth. There are sixteenth-century paintings that feature rainbows with four or even five colours, though no explanations are offered for why the number five might be significant.

What was Shakespeare's view? When Falstaff describes his bruises as being 'all the colours of the rainbow', he's surely not thinking of only three. Perhaps Shakespeare would have reckoned there were four or five rainbow colours. It's just as likely, however, that he would have said 'thousands' or even 'infinite' (a word he was quite fond of), prone as he was to exaggeration for dramatic effect.

Elizabeth is holding a translucent rainbow that
appears to have three bands of colour.

SHAKESPEARE'S COLOURS

When daisies pied and violets blue
And lady=smocks all silver=white
And cuckoo=buds of yellow hue
Do paint the meadows with delight

LOVE'S LABOUR'S LOST

As to *which* colours Shakespeare might have described as appearing in a rainbow, we can perhaps make a guess based on the colours he refers to across his work, as shown in the bar chart below. Green, blue and yellow would be likely candidates.

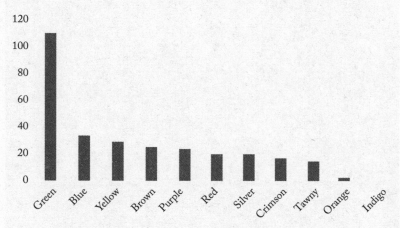

Mentions of colours in Shakespeare's works.

Instead of violet, Shakespeare would most likely have said purple. Shakespeare does mention violet several times, but only in reference to the flower. (Pink was also the name of a flower, while indigo was a dye that was used to give clothing and pottery a dark blue colour.)

Shakespeare might have called the outer colour of a rainbow red, or in a more poetic vein he might have called it crimson.

Of the modern rainbow colours, that leaves only orange. Across all of his works, Shakespeare uses the word orange only twice to refer to colour. In both cases, it is the character Bottom in *A Midsummer Night's Dream* who mentions it. In one scene Bottom is trying to decide what colour of fake beard he should wear when performing on stage. One choice offered to him is an 'orange-tawny' beard. Later he launches into a song that begins:

The ousel cock, so black of hue
With orange=tawny bill.

The ousel cock is an obsolete name for the blackbird, and Shakespeare is describing its beak as orange-tawny, the same colour combination that he used for the beard.

Why the need to add the word tawny? Why not just call the beard and the beak orange? The reason is that orange had not yet become established as a word for a colour. Orange was a citrus fruit, imported mostly from Seville, that was becoming a familiar sight in Tudor street markets. As the bar chart of colours shows, tawny was a popular colour word in Shakespeare's England, a light-brown colour that these days we mainly encounter in reference to the 'tawny owl'. When Shakespeare referred to 'orange-tawny', he was saying that the beard was tawny (everyone knew what colour that was), but similar in hue to one of those fruits that you might see in the market. It's the same as describing something as 'lime green'.

Shakespeare's tentative introduction of the word orange to describe a colour was a big step forward for the artistic palette. Until then, there simply wasn't a word to describe something that today we all recognize immediately.

For example, in the *Nun's Priest's Tale*, one of the Canterbury Tales published around 1380, Geoffrey Chaucer describes a fox's tail as being 'a colour betwixt yellow and red'. 'Dang,' Chaucer must have thought, 'if only there were a word for it!' We still describe people with orange hair as redheads, a holdover from the days when orange wasn't available as a colour description.

But orange didn't yet have a place in the rainbow. As the character Salisbury says in *King John*,

To add another hue unto the rainbow . . . is wasteful and ridiculous excess.

Why put in another colour for the sake of it? As it happens, a hundred years after this was written, that's exactly what happened. And it was a mathematician – Isaac Newton – who gratuitously added not one but two hues to the rainbow, and for unscientific reasons too. You'll find the story behind this in the appendix, on page 190.

If you want to amuse friends with a trivia question, you can ask why Shakespeare never finished a line with the word 'orange'. One reason is that he only used the word six times (twice as a colour, four times as a fruit), and if the word 'orange' was randomly assigned somewhere in a line, there's a good chance it wouldn't have been at the end. Remember, too, that Shakespeare's preference was to finish a line with an iamb, which has a 'dee-dum' rhythm, rather than a word like orange where the stress is on the first syllable. Then of course there's the classic piece of trivia that nothing rhymes with orange – but since most lines in Shakespeare don't rhyme, this particular explanation is a red (or should that be orange?) herring.

THE EMERGING MATH OF COLOUR

Plutus himself . . . hath not in nature's mystery more science than I have in this ring.

ALL'S WELL THAT ENDS WELL

S o far, all this discussion of the number of colours in the rainbow has been rather subjective and philosophical. What about the rigorous mathematics and science behind colours? As it turns out, this knowledge was just taking off.

In late 1605, around the time that *Macbeth* was first being performed, just a few miles from Shakespeare's Globe an English mathematician named Thomas Harriot was making great advances in the study of optics. This 45-year-old polymath had in 1601 become the first person to set out in detail the mathematical laws of refraction, the optical phenomenon that causes white light to split into the different colours.[32] Now he was turning his attention to rainbows.

You have probably never heard of Harriot, and there is a simple reason for this. Although he wrote his theories down in private letters, he never published them. This might have been down to modesty, or perhaps he just wanted to keep his head down, because Harriot had already spent some time in prison. Harriot's patron was Henry Percy, and unfortunately for Harriot, Percy was the cousin of Thomas Percy, one of Guy Fawkes' co-conspirators in the Gunpowder Plot that had been foiled in November of that year. Thomas Percy and Fawkes had already been brutally executed. Anyone who had a personal link to these traitors was under suspicion, and that included Harriot. He was

---

32 Today this is known as Snell's Law, but Snell discovered it twenty years after Harriot.

briefly imprisoned at the Gatehouse prison near Westminster Abbey, before being released without charge.[33]

Thomas Harriot was four years older than Shakespeare. He graduated with a bachelor's degree from his hometown of Oxford in 1580, an education that would have included a good grounding in mathematics. At the same time as Harriot was graduating, the seafaring adventurer Walter Raleigh (not yet a Sir) was making plans to colonize North America. Thanks to a recommendation from the principal of his college, Harriot was hired by Walter Raleigh to improve the navigation skills of Raleigh's ship captains.

Realizing he needed to deepen his knowledge, Harriot immersed himself in the study of the mathematics of navigation, which involved geometry, astronomy, arithmetical calculation and the discipline of precise measurement using tools such as quadrants (a smaller version of the giant protractor that Tycho Brahe had used). Raleigh was impressed by his new recruit, who over the years became his trusted right-hand man. Harriot applied his mathematical knowledge in diverse ways: looking after Raleigh's accounts, conducting land surveys in Virginia and even helping with the design of Raleigh's ships.

When he returned to England, Harriot left his rather chaotic employment under Raleigh and settled into a life of scientific study. Perhaps because of his years observing the sun and stars, and the growing interest in lenses and telescopes, he developed a particular interest in optics. Around 1601, Harriot discovered that different colours of light bend at different angles when they pass through glass, water and other media, and he worked out a mathematical way to measure what we now call the refractive index of red, yellow and green light. (For some reason he didn't get as far as blue.)

---

33 Even Shakespeare had an awkward family link to one of the Gunpowder plotters, Robert Catesby, via his father. This was one reason why he was so keen to write *Macbeth*, a play that gave him a chance to flatter the Scottish King James.

In 1605, fresh out of prison, he used this knowledge to develop his (as it turned out correct) theory for how rainbows are formed when light passes through droplets of water in the sky, and he also established why rainbows have that familiar semicircular shape. He seems to have pretty much cracked everything there was to know about rainbows except for one thing: how many colours there were. As a radical scientific thinker who was challenging the old worldview, Harriot would probably have dismissed the old-fashioned Aristotelian idea of a three-colour rainbow. However, as far as we know, he did not set out a description of all the colours he could see. Maybe he felt no need to do so. Or perhaps, from his examination of the spectrum, he could tell that the colours blended into each other and there were more shades than he could count.

I was hoping that, just as Shakespeare made reference to the astronomical discoveries of Galileo and Brahe, he might also have mentioned Thomas Harriot's breakthroughs in the world of optics. Alas there is nothing in the plays or sonnets that refers to the new science of colour. This is perhaps not surprising given how secretive Harriot was about his discoveries.

However, it turns out that, although Shakespeare may not have known about Harriot, we can be confident that Harriot knew about Shakespeare, and may even have watched him perform.

Harriot was an acquaintance of the playwright and actor Christopher Marlowe, a friend of Shakespeare. They shared atheistic beliefs, and Harriot was even mentioned in a witness's testimony in preparation for Marlowe's trial for atheism. Then there was Harriot's close friendship with Raleigh, who we know attended some of Shakespeare's plays. If Raleigh was going to the theatre, he would surely want to go with a friend, and Harriot would be an obvious choice. At very least, Raleigh would have told his colleague about the plays.

But perhaps the most compelling link between Harriot and Shakespeare arose from the debate at the time about whether or not matter is made up of atoms. The idea of tiny indivisible particles went back to ancient Greece ('a-tom' means cannot be cut). It was a controversial topic, because some of the ideas behind it went against the views of the Christian Church. We don't know if Shakespeare believed in it, but he used the word three times (calling them 'atomies' rather than 'atoms'). For example, in *As You Like It*, Celia says:

It is as easy to count atomies as to resolve the propositions of a lover.

(I.e. it's not easy at all!)

Thomas Harriot himself was a believer in atomic theory (a so-called 'atomist'), but he knew that not everyone agreed with this newfangled view. In a note in one of his manuscripts Harriot joked that the debate about atoms was 'Much ado about nothing'. Now where had he heard that phrase before?

CHAPTER IX

PUTTING INK TO PAPER

WRITING WITH A QUILL

Her maid is gone, and she prepares to write,
First hovering o'er the paper with her quill.

THE RAPE OF LUCRECE

'Shakespeare wrote with a pencil,' goes the old joke, 'but we don't know if it was 2B or not 2B.' Ha ha. That punchline even makes it onto the official Royal Shakespearean Company gift pencil.

Of course, we all know that Shakespeare wrote with a quill. Just about every image and statue of the bard sees him poised, white feather in hand, ready to write. This includes the pork butcher (or accountant) effigy above his grave that I mentioned in Chapter 4, which is now believed to be the truest likeness of him.[34]

And yet – the first rudimentary 'lead' pencils arrived in London for the first time in the early 1600s. We know that at least one of Shakespeare's

---

34 This is according to the research of Lena Cowen Orlin in 2021. Incidentally, the original stone quill in Shakespeare's grave monument, carved in 1520, was stolen, and there is a long-standing tradition that the head pupil at Stratford Grammar School inserts a fresh goose quill into Shakespeare's hand on 23 April – Shakespeare's birthday – each year.

close friends had encountered them. Did Shakespeare himself try out a pencil when writing one of his later plays? Perhaps the joke is not a joke at all. I'll come back to that later.

The quills and pencils of Elizabethan England both raise topics that are intriguingly mathematical: the mechanics of writing with a quill, and the economics of the pencil industry, which was in its infancy during Shakespeare's lifetime.

As far as we know, quills were first used in Spain in the sixth century. They were to remain the writing implement of choice in Europe for over a thousand years. The word 'pen' comes from *penna*, the Latin word for feather. (Penne pasta is so called because it is the shape of a pen nib.)

Quills could be made from the feathers of almost any bird, but by far the most popular were goose feathers,[35] which had the best combination of toughness, internal tube thickness and everyday availability. Every town and village would have had plenty of geese. Not only did they offer a ready supply of free quill feathers (geese moult, like every other bird), but geese were also great egg-providers and they provided the staple meat for the Michaelmas fair in September, and Christmas dinner too, for those who could afford it.

What you might never have thought about, however, is the curvature of a feather. When viewed from below, a feather taken from the right wing of a bird bends round to the left, while a feather taken from the left wing bends to the right.

For the quill user, this has practical consequences. If you are right-handed, a right-wing feather bends towards your nose and obscures your view of what you are writing, which is an irritating distraction. A left-wing feather, on the other hand, bends pleasingly away from your nose leaving you with a clear line of sight.[36] Since Shakespeare

---

35 The quote from *Hamlet* 'Many wearing rapiers [swords] are afraid of goosequills' probably inspired the modern expression 'The pen is mightier than the sword.'

36 In practice, quill users would often strip the 'fluff' off the quill, leaving a stem that looked similar to a modern pen.

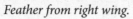

Feather from right wing.   *Feather from left wing.*

was right-handed, he must therefore have been a left-wing writer. This revelation will no doubt provoke consternation among some readers of the *Daily Mail*.

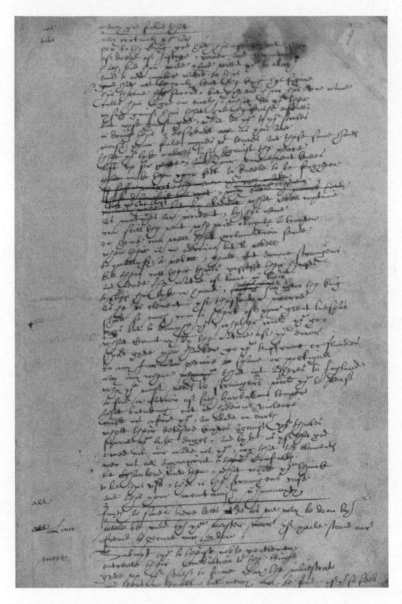

A page from the play Sir Thomas More, *which academics are confident is written in Shakespeare's hand. Note the crossings out. The slope of the letters confirms he was right-handed.*

WRITING WITH INK

O, none, unless this miracle have might,
That in black ink my love may still shine bright.

SONNET 65

Ink was made from a curious concoction of iron sulphate (known at the time as vitriol), crushed oak galls, gum arabic and water. The iron sulphate probably came from Kent, and was produced from iron pyrite, the shiny mineral better known as fool's gold ('all that glisters is not gold', as Shakespeare says), which reacts with air to produce the valuable sulphate form. Oak galls are the apple-like balls that grow on many species of oak tree. Gum arabic is the sap from acacia trees that grew in Africa and the Middle East.[37] The iron sulphate reacts with the tannic acid in the galls to create iron tannate, which turns black when exposed to air, and the gum thickens the fluid to make it usable as ink. The Elizabethans had no idea about the detailed chemistry of all this, but ink manufacture was an early example of a chemical industry that relied on long-distance trade.

As well as being an ink ingredient, the word 'gall' also meant anger or spitefulness. The ink ingredient is being deliberately used as a pun when, in *Twelfth Night*, Sir Toby Belch says:

Let there be gall enough in thy ink, though thou write with a goose-pen, no matter.

Shakespeare probably bought his ink and paper from a book-selling stationer, maybe one of the many based around St Paul's Cathedral, but it's possible that he bought the ingredients to make ink himself, as many people did. Making good ink meant putting in

---

37 'Drop tears as fast as the Arabian trees their medicinal gum' (*Othello*).

the correct ratio of ingredients, so some basic numeracy was needed for ink-making, just as it was (and still is) for baking a cake. There were plenty of books with ink recipes, including a 1571 manual by John de Beau Chesne and John Baildon who put their instructions in rhyme. (I've modernized the spelling.)

To make common ink of wine take a quart,
Two ounces of gum, let that be a part,
Five ounces of galls, of copres[38] take three,
Long standing doth make it better to be;
If wine ye do want, rain water is best,
And as much stuff as above at the least . . .

Writing involved dipping the quill into the inkpot, or inkwell, and scratching out perhaps eight or ten words before having to re-dip the nib when the ink ran out. After writing a few pages, the nib of the quill would wear out, and would therefore need regular sharpening with a penknife, a device we still use today though not usually for sharpening pens.

You might think that a wordsmith like Shakespeare would find the process of writing with a quill painfully slow, but in fact an experienced writer could rattle off text at 20–30 words per minute, which is not that much slower than the word rate of an untrained modern typist.

There were some shortcuts to speed the process up. My favourite example is the way that the Roman numeral for ten, X, was written not with two strokes of the pen, but with a single loop that resembles a piece of ribbon. The image on the next page shows how the theatre manager Philip Henslowe wrote the number XX (Roman numerals for 20) in his accounts from 1595.

---

38 'Copres' is copper sulphate, which would result in a brown ink rather than the black you got with iron sulphate.

The number XX (twenty) as written by Philip Henslowe in his accounts.
Note how each X is produced in a single swirl of the quill, either for speed
or to reduce the risk of an ink blot. They barely resemble Xs at all.

Despite the writing speed that could be achieved with a quill, it would have been frustrating to work with such messy, indelible materials. You can see why the market was ready for a writing implement that was cleaner and required fewer ingredients. And how much neater it would be if the errors could be erased rather than crossed out. In other words, exponents of the written word were ready for the pencil.

ARRIVAL OF THE PENCIL

. . . the fisher with his pencil and the painter with his nets.[39]

ROMEO AND JULIET

S hakespeare certainly knew about pencils: in *Romeo and Juliet* the servant refers to one, and there are mentions in several other plays and sonnets. However, the object that Shakespeare called a pencil was actually a fine brush used by artists. But, as fate would have it, when *Romeo and Juliet* was being performed in London, a discovery in the north of England had already laid the foundations for the production of the modern pencil.

By a remarkable coincidence, in the year that Shakespeare was born, 1564, at least one significant event (perhaps two) happened that would ultimately lead to the pencil industry, in which England was to become the world leader.

Some histories of Cumbria claim that it was in 1564 that, halfway up the side of a valley in Borrowdale in the Lake District, an ash tree blew over. Tangled in its roots were some black rocks. The rocks were shiny like metal, but felt slippery to the touch and left black marks on your hand. The metallic appearance of this odd substance meant that some people called it lead, or 'plumbago', even though it had no metal in it at all. Locals preferred to call it 'wad'.

This substance had been known of in Cumbria for many years. It could sometimes be found lying loose among the stones, and shepherds found it handy for marking sheep. But the lumps in the tree roots were, according to folklore, the most concentrated find they'd ever seen, and

---

39 This is a comic moment in which a servant gets confused. A painter would use a pencil, a fisher would use a net.

a clue to the fact that there was a rich seam of this material lying just beneath the surface.

Whether the tree really did fall in the year of Shakespeare's birth can't be verified, but we do know that it was in 1564 that Queen Elizabeth I initiated the Company of Mines Royal. The purpose of this company was to explore for – and then mine – highly prized copper, gold, silver and (metallic) lead ore, from which the queen would, she hoped, earn huge royalties. Experienced miners were brought in from Bavaria and sent to the Lake District, where it was known that there were likely to be plentiful supplies of lead and copper, in particular. Little did they suspect that the most valuable resource would prove to be a different kind of 'lead', the pencil variety, which would not acquire its modern name of graphite until the nineteenth century.

The Company of Mines Royal was what was known as a 'joint stock company', which meant that people could buy and sell shares in it. The investors were typically wealthy gentry, entrepreneurs and merchants (Shakespeare was not one of them – his business interests were rather more parochial, and he might not even have been aware of the burgeoning mining industry – he never mentions it).

This method of financing made it much easier to raise capital, and there are some who argue that this, together with the commercial use of expert miners, made the mining industry of windswept Borrowdale in 1600 the true start of the Industrial Revolution. The company fuelled the need for Elizabethan accountants and financiers too.

To start with, the Bavarian miners concentrated on copper mining not far from the town of Keswick, and little attention was paid to the wad. Meanwhile, local miners in Borrowdale dug down from the surface to extract wad from above, but they found that holes they dug would quickly fill with water. What was needed was good drainage, and the Bavarian miners, who were now well established in copper mining, were brought in to solve the problem. Their solution, at some time in the early 1600s, was to dig horizontal tunnels into the hillside until they met the vertical seam of wad. Remarkably, it is still possible to walk through these 'German' tunnels.

Mark Hatton, a mine historian in Cumbria, kindly agreed to give me a guided tour. Kitted in helmets and kneepads, we walked up the steep valleyside above the village of Seathwaite, and Mark quickly located one of the tunnel entrances. I asked if the dank, low-ceilinged tunnel that we were in was one that had been dug in Shakespeare's lifetime. Alas, it wasn't, the original German tunnel was a few feet above us and to reach it we would have needed to do some more serious potholing. I lost my nerve at that point. Fortunately, Mark was able to show me a photo of the original tunnel. The passage is straight and its walls are beautifully symmetrical, with what is described as a coffin-shaped cross-section. It's a remarkable bit of engineering, and it's incredible to think that these shafts and tunnels were being chipped out using rudimentary Elizabethan tools, while Shakespeare was busily writing plays with a quill three hundred miles away in London.

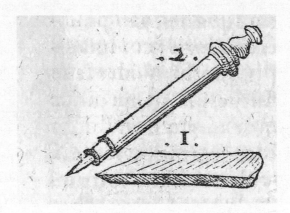

The sketch of a prototype pencil done by inventor Conrad Gessner of Switzerland in 1565, when Shakespeare was one year old. Was it ever made? Did it use Borrowdale graphite?

The price per ton of pencil lead in 1600 was about £15, roughly the same as a ton of metal lead, which certainly made it worth mining. A pebble-sized lump, big enough to write with, would probably have cost less than a penny, even after the costs of transporting it. But nobody had yet grasped just how valuable a lump of graphite would become. Although it quickly became a prized resource for writing, it proved to be of far higher value as a lubricant for moving parts in a machine and, crucially, as a mould for manufacturing perfectly smooth, spherical cannon balls. By the time of the Napoleonic Wars in 1812, Borrowdale graphite was in such demand that its price had rocketed to £3500 per ton, not far off the value of silver. The owners of the wad mine would become phenomenally rich.

The first pencil factory in Cumberland wasn't opened until 1832, but in Shakespeare's time people were already finding ways to improvise homemade graphite pencils.

By 1610, when Shakespeare was writing *Cymbeline*, 'black lead' was being sold in the streets in London. To make a pencil, a stick of wad would be wrapped in paper, string or twigs. Anyone using a pencil probably didn't try to sharpen the tip, but if they did, the black crumbs would be worth saving as a lubricant for, say, a wheel or a rusty hinge.

These new writing instruments were being used in school too. One of the first teacher-training books was written by John Brinsley in 1612. His book was entitled *Ludus Literarius, or The Grammar Schoole*. The blurb described it as 'shewing how to proceed from the first entrance into learning to the highest perfection required in the Grammar Schooles'.

In it, Brinsley set out advice on everything a teacher might want to know about good teaching practice. He went into meticulous detail on how to teach Latin and its grammar, but he also included much broader advice, such as tips on how to keep discipline in the classroom. Of most interest to us here, however, is his recommendation of best practice when pupils were writing out their work.

His advice was that scholars should write their neat finished work in ink, but that for their rough notes and first drafts he recommended using a pencil 'of black lead'.

It is best to note all schoole books with inke ... because ink will indure ... But for all other bookes, which you would have faire again at your pleasure; note them with a pensil of black lead.

The advantage of writing with a pencil, as Brinsley knew, was that you could rub it out. But what with? Rubbers (or erasers, as Americans call them) would not be invented for another 150 years. As Brinsley explained, a Shakespearean eraser would have been made of a more familiar everyday material:

For that you may rub out againe when you will, with the crums of new wheate bread.

There's evidence that lead pencils were becoming a part of a mathematician's toolkit at this time – not from Shakespeare, but from his friend and fellow playwright Ben Jonson. In his 1609 play *Epicoene*, Jonson's character Sir Amorous la Foole describes a mathematician's equipment:

For the mathematics: his square, his compasses, his brass pens, and black-lead, to draw maps of every place and person where he comes.

Black lead for sketching maps – so why not for writing, too?

If Shakespeare did use a pencil when making tweaks to his final play, it might have looked like the stick of graphite shown here, wrapped in string. And he'd have had a slice of bread to hand, for when he needed to make corrections.

An early Jacobean pencil, with bread used as a 'rubber'.

There is one final thing to bear in mind. If Shakespeare did ever use a pencil, it would have been made from pure Borrowdale wad, which was the finest natural graphite that has ever been found in the world.

2B or not 2B? It would have been a 2B at very least, and possibly a 9B, the softest, purest pencil of all.

The mining tunnel dug to reach the seam of wad (around 1610).

PRINTING AND PUBLISHING

Devise, wit; write, pen;
for I am for whole volumes in folio

LOVE'S LABOUR'S LOST

Handwritten plays were fine for actors but, for wider circulation, the works needed to be printed. Around 1450, Johannes Gutenberg had invented the printing press, and within thirty years, William Caxton had introduced printing presses to England. By Shakespeare's time, printed pamphlets and books were widespread. With the printing industry came a whole new application for numbers. About half of Shakespeare's plays and all of his sonnets were printed in his lifetime. It's believed that he had little if any control over their production, apart from his poems *Venus and Adonis* and *The Rape of Lucrece*, for which he might even have done the proof-reading.

The standard format for printing books at that time was known as quarto. Eight pages were printed on a single sheet (four on each side), in the layout shown in the diagram opposite. Once printed, the sheet was folded first along the horizontal line, then the vertical, so that the eight pages were lined up in the correct order, the right way up. If you have a blank sheet of paper you can have a go at reproducing it. Write the numbers 1 to 8 on front and back, making sure you put them the correct way up. If you get the folding right, you should find that the numbers line up in order front to back, with each number the correct way up.

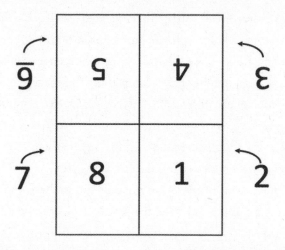

Numbering of quarto pages. Page 2 goes on the back of 1, 3 is upside down on the back of 4, etc. To create eight quarto pages in the correct order, fold along the horizontal centre line, so that 4 is behind 1, then along the vertical centre line.

To create a book, several of these eight-page segments would be stitched together. The folded edges that weren't part of the spine were then cut so that the pages could be turned and read.

It would have been easy for a novice printer to make mistakes. They might create pages whose reverse was upside down, or pages that were in the wrong order, but if such errors happened, they have not survived to the modern day. Perhaps those imperfect editions were pulped, or simply not looked after and disposed of when they were worn. Most of the quartos, including the first edition of Shakespeare's Sonnets, didn't have page numbers. At the bottom of every other page there was a code such as D3, but this was for the printer's benefit as it indicated which sheet the page belonged to: there would be an A sheet, a B sheet and so on, each folded to make four leaves printed on both sides to make eight pages. If a page had D3 at the bottom, that meant that it was the third (of four) leaves from sheet D.

In 1623, seven years after Shakespeare's death, his friends put together the First Folio, a much more substantial book, a collected edition of all of his plays. This time there were page numbers – and, sure enough, page number errors slipped in. The most glaring error in the First Folio appears in *Hamlet*. The page after 157 is numbered 258, and the numbering continues upwards from there. An easy mistake to make, and a very time-consuming one to correct. There were numerous other errors too – words squeezed in to fit the page, large blank spaces padded out – because there were none of the modern tools of being able to press a button and have the words paginate and fill out lines automatically.

The print quality of Shakespeare's first published work, *Venus and Adonis*, was considerably more impressive. This narrative poem was published in 1593. The printer was Richard Field, originally of Stratford-upon-Avon, and probably a friend of Shakespeare's since childhood.[40]

I was curious to see what sort of page numbering was used in 1593, thirty years before the First Folio, and tracked down a book that claimed to be a facsimile of the original edition. In this book, not only are all the pages numbered, but the lines of verse are numbered too, albeit in an unusual way.

Every verse has six lines, which rhyme in the pattern ABABCC. The first, fourth and sixth lines are numbered, but the other three are not. The first verse has lines 1, 4 and 6 numbered in the margin, and the numbering continues up from this, so that if you turn to, say, page 26, in the left column next to the top verse are the numbers 547, 550, 552, and for the next verse, 553, 556, 558 (see the image on the next page).

---

40 The Field and Shakespeare families lived in adjacent streets in Stratford and were both in the leather industry – tanning and glove-making, respectively. Richard was only a couple of years older than William, so it's almost certain that they saw each other every day in the school classroom and probably in the street too.

26

	VENVS AND ADONIS.
547	Now quicke defire hath caught the yeelding pray,
	And gluttonlike fhe feeds, yet neuer filleth,
	Her lips are conquerers, his lips obay,
550	Paying what ranfome the infulter willeth:
	VVhofe vultur thought doth pitch the price fo hie,
552	That fhe will draw his lips rich treafure drie.

Facsimile of the first quarto of Venus and Adonis,
1593, from the Bodleian Library, Oxford.

The numbering is immaculate right through to the final line, number 1194 ... except for a typo on the second page. The first line at the top of the page should be numbered 19, but a rogue extra 4 has been set, making the first line 194. Everything else is correct. Had I discovered that even the meticulous typesetting of Richard Field was flawed? No. It turns out that this numbered edition was produced in the nineteenth century.

There was no page numbering in the original edition of *Venus and Adonis*. Page numbering would not become the norm for several more years. In fact the only numbers to appear in that book were the date on the title page, which is in Arabic numerals, and the printer's quarto sheet numbering at the bottom of every fourth page. Instead of B3 it says B iij, a Roman 3. The 'j' at the end is not a typo. It was the custom when writing Roman numerals to write the final 'i' as a 'j', so that nobody could add another digit at the end.

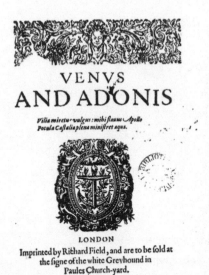

VENVS AND ADONIS.

Neuer did paſſenger in ſommers heat,
More thirſt for drinke, then ſhe for this good turne,
Her helpe ſhe ſees, but helpe ſhe cannot get,
She bathes in water, yet her fire muſt burne:
Oh pitie gan ſhe crie, ſlint-hearted boy,
Tis but a kiſſe I begge, why art thou coy?

I haue bene wooed as I intreat thee now,
Euen by the ſterne, and direfull god of warre,
VVhoſe ſinowie necke in battell nere did bow,
VVho conquers where he comes in euerie iarre,
Yet hath he bene my captiue, and my ſlaue,
And begd for that which thou vnaskt ſhalt haue.

Ouer my Altars hath he hong his launce,
His battred ſhield, his vncontrolled creſt,
And for my ſake hath learnd to ſport, and daunce,
To toy, to wanton, dallie, ſmile, and ieſt,
Scorning his churliſh drumme, and enſigne red,
Making my armes his field, his tent my bed.

Thus he that ouer-ruld, I ouer-ſwayed,
Leading him priſoner in a red roſe chaine,
Strong-temperd ſteele his ſtronger ſtrength obayed.
Yet was he ſeruile to my coy diſdaine,
Oh be not proud, nor brag not of thy might,
For maiſtring her that foyld the god of fight.
B iij

VENVS
AND ADONIS

Vilia miretur vulgus: mihi flauus Apollo
Pocula Caſtalia plena miniſtret aqua.

LONDON
Imprinted by Richard Field, and are to be ſold at
the ſigne of the white Greyhound in
Paules Church-yard.
1593.

Pages from the first edition of Venus and Adonis. The publication date is in Arabic numerals, but the quarto numbering uses Roman numerals.

By the end of Shakespeare's writing career, the switch from Roman to Arabic as the standard numbering system was almost complete.[41] In the first edition of his *Sonnets* (published in 1609), the individual sonnet numbers and the printer's quarto numbers at the foot of the page are all Arabic numerals.

---

41 The gradual switch from Roman to Arabic numerals can also be seen in Philip Henslowe's accounts for the Globe theatre – see the appendix on page 193.

CHAPTER X

MATHEMATICS, MAGIC AND WITCHCRAFT

JOHN DEE, MATHEMATICIAN AND MAGICIAN

What black magician conjures up this fiend,
To stop devoted charitable deeds?

RICHARD III

Mathematics and magic have had a long association. Many magicians are mathematicians, and vice versa. These days we tend to associate magic with entertainment and trickery. In Elizabethan times, however, magic usually had much darker connotations. It covered anything that couldn't be explained, so many chemical and physical phenomena were deemed to be magical. Mathematics, with its mysterious symbols, was also often treated with suspicion, and in the eyes of some it was dangerous magic.[42]

---

42 In his 1690 book *Brief Lives* John Aubrey comments that in Elizabethan times 'Astrologer, Mathematician and Conjurer were the same thing.'

The word 'calculating' was synonymous with 'conjuring'. In *Julius Caesar,* when Cassius wonders 'Why children calculate' he's not commenting on their school arithmetic but is instead pointing out that children have mysteriously been making prophecies (one of the signs that a 'monstrous' government is soon to be in charge of Rome).

Magic could be for the good ('white magic'), but it could also be sinister ('black magic'). At its most benevolent, magic was used to tell people's fortunes and to heal. At its worst, it was witchcraft.

In Shakespeare's England, the mathematician most associated with magic was John Dee. A prodigious scholar, he was appointed as a fellow of the newly founded Trinity College Cambridge in 1546 at the age of just nineteen. He was a reader in Greek but his passion was mathematics and natural philosophy, and he was determined to use his knowledge to understand the secrets of the universe. Long before Harry Potter, John Dee was on a quest to find the mythical Philosopher's Stone – a substance with which you could achieve immortality and turn base metals into gold – and he believed mathematics was part of the secret to finding it.

The first act that drew widespread attention to John Dee was when he put on a production of the ancient Greek play *Peace* in his college's main hall. At one point in the play, the lead character Trygaeus wants to consult the God Zeus. To reach the heavens, Trygaeus flies on a giant beetle. In ancient Greece, perhaps this 'flight' had been left to the audience's imagination, but John Dee set about building a mechanical device (the details of this remain a mystery) that enabled Tyrgaeus, mounted on the beetle, to soar up to the ceiling of the hall. The audience gasped in astonishment, and this dramatic staging effect was to remain a talking point for many years. Whatever the mechanics behind the beetle was, John Dee would have regarded it as mathematics.

Dee remained fascinated by physical stunts like this, and later in life, visitors to his house would sometimes be treated to amazing effects based on such unexplained phenomena as magnetism and chemical reactions – dropping iron filings into a flame to create a sparkler effect, for example. Dee's purpose was not to 'trick' people, however, but rather to share with them the wonders of nature. In 1672, an elderly woman called Goodwife Faldo recalled a moment from her childhood when she (aged six) had been invited with her mother into John Dee's home. He had taken them to a darkened room and shown them an image of a solar eclipse, projected through a pinhole. It must have been an awe-inspiring and magical experience for a young girl, and probably for her mother too.

Although there's no evidence that John Dee and Shakespeare ever met, many believe that Dee was the inspiration for the character Prospero in *The Tempest*. Prospero is a magician who uses his magical powers to govern everything that happens on his island. Prospero's servant is Ariel, who also performs magic. Early in the play Ariel helps Prospero to conjure up a storm ('tempest') to shipwreck Antonio, and later on he performs a couple of magical stunts which include making a banquet disappear. The staging instructions say

> *Thunder and lightning. Enter ARIEL, like a harpy; claps his wings upon the table; and with a quaint device the banquet vanishes*

There are no clues as to what this 'quaint device' is, but presumably something mechanical. It's reminiscent of Dee's flying beetle, if not quite so spectacular.

John Dee and Prospero also shared a love of books. Prospero tells of how he was banished from Milan and cast off in a small boat, but was able to take with him a few of the most precious books from his library, thanks to a kindly nobleman.

> *Knowing I loved my books, he furnish'd me From mine own library with volumes that I prize above my dukedom.*

John Dee's library at his house in Mortlake on the River Thames was famous. He claimed to have over 3000 volumes, on subjects that included mathematics, natural history, music, astronomy, cryptography, ancient history and alchemy. Sadly, much of that collection was sold or stolen while Dee was travelling abroad, and most of the books were never recovered.

We don't know what John Dee's most prized book was, but, for Prospero, one book seemed to have more importance than any other: his book of magic. He refers to it before performing a number of magical acts during the play,

> I'll to my book,
> For yet ere supper-time must I perform
> Much business . . .

and towards the end, with his work done, he announces that he will throw his book into the sea.

> But this rough magic
> I here abjure . . .
> I'll drown my book.

Where did Shakespeare get the idea for a magic book? Perhaps he was thinking of a book that had been published when he was a young man, and which we know influenced his thoughts about magic and, in particular, witches.

John Dee (1527–1609), mathematician, astrologer and alchemist, who became scientific adviser to the queen.

THE BOOK OF MAGIC

How now, you secret, black and midnight hags?

MACBETH

O ur modern image of a witch as a haggard, cackling old woman owes much to the three witches of *Macbeth*. Yet Shakespeare himself almost certainly got the idea for witches such as these from an astonishing book entitled *The Discoverie of Witchcraft*, written by Reginald Scot in 1583.

Shakespeare's life coincided with a huge escalation in witch-hunting in England and Scotland. It was widely believed that witches had the power to hurt or even kill people, and if somebody had a severe change of fortune or became seriously ill, it was common to accuse a person they suspected – or had a grudge against – of witchcraft. Henry VIII's Witchcraft Act of 1541 was the first time that witchcraft became officially punishable by death. Another Act followed in 1562 under Elizabeth I, and in 1603 a more draconian one under King James I. James was particularly obsessed with witches – he believed them to be one of the greatest evils known to mankind and that they had to be eradicated. The persecution of witches rose considerably under his reign. He even wrote a book about them, *Daemonologie*, in which he gave detailed descriptions of their craft. In 1612, not long after *The Tempest* was first performed, eight men and two women from Pendle in Lancashire were famously found guilty of witchcraft and hanged.

Given the mood of the times, you might expect that a book called *The Discoverie of Witchcraft* would be cashing in on the hysteria, as some sort of witch-finder manual written to help people identify potential witches. Remarkably, the aim of the book was the complete opposite. Its author was a whistleblower, a sceptic who wanted people to be objective, and to see that most so-called witchcraft was nothing of the kind.

Reginald Scot was a well-educated country gentleman with strong Protestant views. The view among Calvinists like him was that witchcraft was mere superstition, and those who believed in it were turning their backs on God. As Scot said: 'it is neither a witch, nor devil, but glorious God that maketh the thunder ... God maketh the blustering tempests and whirlwinds ...'

Scot therefore set about writing a manual to explain that most so-called witchcraft was simply conjuring tricks and sleights-of-hand, and that accusations of witchcraft were often no more than prejudice against innocent and vulnerable old women (though witches could be young and even male too). This was a book that advocated rational thinking based on evidence. Needless to say, the superstitious King James hated the book, and it's sometimes claimed that he tried to have all copies destroyed. If so, then he failed.

Before Scot's book was published, there was no particular image of what a witch looked like. Scot changed that by offering a very vivid description:

One sort of such as are said to be witches, are women which be commonly old, lame, bleare-eyed, pale, foul, and full of wrinkles.

This description of witches was a radical shift in the way they had been portrayed in previous versions of the Macbeth story, and it is widely believed that Shakespeare's witches in *Macbeth* were influenced by Scot's description.

What are these
So wither'd and so wild in their attire,
That look not like the inhabitants o' the earth,
... By each at once her chappy finger laying
Upon her skinny lips

Controversial books are always popular, so Scot's book was widely read. And anyone who read it would have learned about a lot more than just witchcraft, because this was also the first book to explain in detail how to perform conjuring tricks.

On one page, Scot explains how to get a 'decapitated' head to talk, by using a box with a secret hole in it. The accomplice hides in the box with his head sticking out of the hole through a plate with a false bottom. A gullible audience sees a seemingly disembodied head resting on a plate, which then opens its eyes and starts to talk.

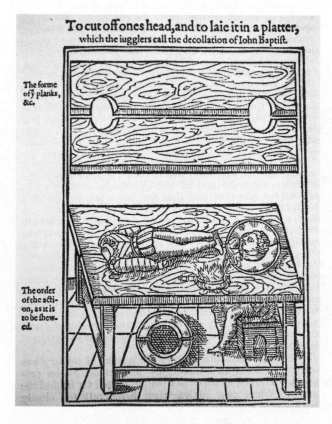

Illustration from Reginald Scot's Discoverie of Witchcraft, *showing how to create the illusion of a talking decapitated body.*

There is also a popular magic trick called 'fast or loose' which Reginald Scot describes, and which Shakespeare mentions three times. The version that Scot explains involves placing three beads onto a cord, and then removing them while the two ends of the cord are being firmly held by a volunteer. Shakespeare refers to the trick as 'fast AND loose' so it's possible that he is referring to a different trick by that name that was popular at fairs. The most obvious reference to fast and loose is in *Antony and Cleopatra*:

Like a right gipsy, hath, at fast and loose, Beguiled me to the very heart of loss.

Fast and loose was presented as a game. A belt or cord was folded in half and then wound into a coil as shown in the diagram. A contestant had a choice of where to place their finger in the coil, A or B.

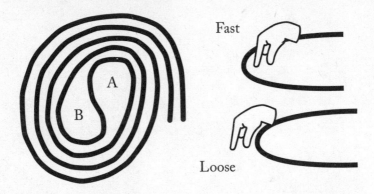

Once they had made their choice, the ends of the belt were pulled, and if the finger ended up pinning the belt, 'holding it fast', then the contestant won, but if the belt was freed from the finger ('loose') the contestant lost. It might have felt like a fair game with a 50–50 chance of winning if you simply guessed, but the gamester was in fact completely in control of the outcome. If the player chose B, the two ends of the belt were pulled together to release the finger, and if they chose A, the outer coil was first unfurled one rotation anticlockwise and then the ends were then pulled together to achieve the same result.

To know which way to pull the belt, the trickster counts how many layers of belt are between the victim's finger and the outside of the coil – if it is an even number, pull the belt ends together, and if an odd number, unwrap anticlockwise. (I strongly recommend that you find a belt and try this out for yourself.) The modern expression 'playing fast and loose', meaning to bend the rules, almost certainly originates from this game.

Scot also dedicates two pages to conjuring tricks using playing cards. As we saw in Chapter 3, card games were becoming increasingly popular in Tudor times, thanks to cheaper manufacturing. Soon after the arrival of card games came the new entertainment of card *tricks*.

'Pick a card, any card.' Card magic began soon after the arrival of playing cards. Note that cards featured the modern suits clubs, hearts, spades and diamonds but did not have printed numbers. Perhaps the manufacturers weren't sure whether to use Roman or Arabic numerals so they decided to have neither.

Here's one simple card trick that Scot describes. You take a secret peek at a card, shuffle it to the bottom of the pack, and then show the bottom card to your volunteer with your eyes turned away . . .

> willing them to remember it: then shuffle the cards, or let anie other shuffle them; for you know the card alreadie, and therefore may at anie time tell them what card they saw: that will drawe the action into the greater admiration. Neverthelesse would be done with great circumstance and shew of difficultie.

In other words, peek at the card and then after some misdirection, reveal what it is. It's pretty basic, but if you've never seen a card trick before, you might easily fall for it.

Many modern card tricks rely on card counting and the mathematical properties of shuffling and dealing, but Scot makes no mention of this type of trick, perhaps because the patterns behind such tricks hadn't been discovered yet.

THINK OF A NUMBER

I am thinking, brother, of a prediction I read this other day
KING LEAR

For all this talk of the connections between magic and mathematics in Shakespeare's time, I was a little disappointed that none of the magic of John Dee or Reginald Scot included what I would regard as a proper *mathematical* trick. You know the sort I mean: 'Think of a number . . . double it . . . add six . . . divide by two . . . take away the number you first thought of and you have finished on the number . . . THREE.' Would Shakespeare ever have encountered this sort of number trick?

I was beginning to think that mathematical 'mind-reading' must have appeared in later centuries, but in researching John Dee, I came across an unexpected connection to a mathematician that we've met before. In 1561, Dee was asked to revise and add material to a new edition of Robert Recorde's book of arithmetic, *The Ground of Arts*, featured in Chapter 2. It turns out that Dee wasn't the last person to add a chapter in this way. In 1582, a schoolteacher by the name of John Mellis took over the editorship. Mellis added a new Chapter 18 to Recorde's book, with the tantalizing title: 'Sportes and Pastimes done by Number', which I assumed would be about the mathematics of games and sports. Was this perhaps an introduction to the scoring system for tennis, or the rules for card games like Primero?

No. Instead, it turned out to be a series of mind-reading number tricks to be performed on your friends. Several of them are 'think of a number' style tricks (or, as Mellis put it, 'How to know the number that any man doth think, or imagine in hys minde').

My favourite is a simple trick that involves the 'Divination of a number upon the casting of two dice'. Here are the instructions.

(You might want to get a couple of dice and a volunteer helper before you go any further.)

1. Ask your volunteer to roll the two dice and secretly note the total without you seeing them.
2. Still hiding the two dice, ask them to pick up one of them and peek at the number on the bottom, then add this to the total.
3. Next, they should roll the die they just peeked at, and add its number to the running total. They can now reveal the two dice.

Point out that you have no idea which dice they rolled or what number was originally on one of the dice. Suppose the numbers you can now see on the top of the dice add up to 8. With a drumroll you reveal that the total that your volunteer ended up with is . . . 15. Gasps all round!

How does it work? Since Roman times it has been the convention that the opposite sides of a die always add up to seven: 1 is always opposite 6, 2 is opposite 5 and 3 is opposite 4. Nobody knows for certain why this is, but the number 7's mystical connections may have played a part. It means that when you roll a die and add the numbers on the top and bottom, you will always get 7 no matter which number comes up. To perform the trick, all you have to do is add 7 to the two numbers that are showing after the final roll.

Shakespeare would have left school by the time this trick appeared in Recorde's book, but the new chapter would surely have been known to someone among his peers. Picture the scene. Shakespeare and his friends have gathered at the White Hart Inn. They have already had their first ale, and it's time for a game of Hazard. 'But first, gentlemen,' one of them announces, 'I wager that I can read the mind of any fellow that rolls these two dice.'

CHAPTER XI

CODES AND CONSPIRACY

SECRET MESSAGES AND CHRONOGRAMS

Madam, I have a secret to reveal.

HENRY VI PART 1

Shakespeare lived in a world of intrigue. With religious factions at each others' throats and conspiracy everywhere, ciphers and hidden messages were increasingly significant in the lives of the famous and powerful.

On the 17th July 1586, Mary Queen of Scots, living under house arrest in Staffordshire, sent a letter to Anthony Babington. The letter was written in code, and included an instruction that would seal her fate: 'The six gentlemen must be set to work'. This was Mary's tacit approval of a plot in which a group of Catholics were to kill Queen Elizabeth I.

Unfortunately for Mary, the letter was intercepted by a team of code-breakers led by Elizabeth's principal secretary and chief spymaster, Francis Walsingham. Walsingham was tapping into a new mathematical science that was advancing rapidly across Europe, what we now call cryptography. It was the final straw that would lead to Mary's execution the following year.

Not all hidden messages had evil intent, however. Across Europe, there was a popular practice of hiding dates above buildings, in books and in portraits – often just for the fun of it. These hidden dates are known as chronograms, and one example appears in an engraving of Elizabeth (see overleaf), who died in 1603.

In the top left, next to Elizabeth's shoulder, is the inscription *Mortua anno MIserICorDIae*, meaning 'died in the year of mercy'.

Mortua anno
MIserICorDIæ.

Why was 1603 a year of 'mercy'? Perhaps the death of Elizabeth was a reminder to all Christian people that God had the power to take life as well as give it, and that in his infinite mercy he would (they hoped) take the queen to heaven. More likely, however, the engraver chose the word *Misericordiae* because it conveniently contained the letters for a perfect chronogram. Notice how the letters MIICDI have been deliberately capitalized by the engraver. Rearranged, they make MDCIII, 1603, the year of Elizabeth's death. If Elizabeth had died in any other year, the engraver would have needed to find another word or words that had the required number of Is or Vs.

It was around this time that coded messages also began to capture the public's imagination, and fed the notion that perhaps there were secrets to be found in even the most innocent of documents. Eventually the suspicion of hidden messages would spread to the work of Shakespeare himself.

Portrait of Queen Elizabeth I of England, c.1603 (the year of her death), by Crispin de Passe.

Are there hidden dates to be found within the great man's work?

Let's take a closer look at *Macbeth*. Scholars can't agree on which year Macbeth was written. Was it 1606 or 1607? And yet for those who know where to look, the answer is hiding in plain sight.

The top section of the title page of the first edition of Shakespeare's Macbeth.

In the top section of the title page of the play, page 131 (as printed in the First Folio), embedded within the title, are four Roman numerals, D, I, M, C. It's easier to spot them if they are enlarged, like this:

THE TRAGE**DI**E OF **MAC**BETH

Rearranged into descending order they form the number MDCI (1601). Add to this the three other numerals 1–3–1 that form the page number and you get 1606.

But why would the printers go to all that trouble just to hide the date of the play?

You are right to be sceptical. This revelation about the date of the play is my own invention, created after spending an hour or two searching through the First Folio for numbers that might add up to a date that had some significance in the writing of one of Shakespeare's plays. I've done this to show that it doesn't take much effort to find messages when there are none.[43]

Chronograms might have been popular in continental Europe, but they were very rare in England. Few scholars believe that there was an agenda among the printers or anyone else to hide dates within *Macbeth* or indeed any other play. And yet there was, and still is, a huge interest in numerology that seeks to find hidden patterns and secret messages in the work of Shakespeare, based around numbers. Most of this is used to back up the arguments of the sceptics who believe that Shakespeare was not the true author of all the plays and poems published under his name. There are broader claims, too, that hidden in his language there are coded messages that he was a closet Catholic, that he translated passages in the King James' Bible, that he smoked marijuana, and I even read one claim that Shakespeare predicted the assassination of Abraham Lincoln.

In the spirit of the debunker of witchcraft, Reginald Scot, we need to be sceptical about some of the outlandish claims that are made on the basis of mathematical, or pseudo-mathematical, analysis of Shakespeare's work.

---

43 Another reason to be sceptical is that chronograms in that era only used Roman numerals, and the rest of the text was usually in lower case so that the capital letters would stand out.

ACROSTICS AND OTHER
HIDDEN WORDS

Yea the illiterate, that know not how
To cipher what is writ in learned books,

THE RAPE OF LUCRECE

A form of secretive wordplay that would have been more familiar to Shakespeare was the acrostic. This device spells out a word or message using the first letter of each line of a poem or piece of text.

Acrostics were used occasionally by the ancient Greek and Roman poets, and they were still popular in the 1600s. In 1606 Ben Jonson wrote the play *Volpone*, which premiered at the Globe theatre. *Volpone* (meaning 'The Fox') is a tale of greed and punishment and the printed version opens with a verse that summarizes the plot:

Volpone, childless, rich, feigns sick, despairs,
Offers his state to hopes of several heirs,
Lies languishing: his parasite receives
Presents of all, assures, deludes; then weaves
Other cross plots, which ope themselves, are told.
New tricks for safety are sought; they thrive: when bold,
Each tempts the other again, and all are sold.

The capitalized letters that start each line spell V-O-L-P-O-N-E. Let's be clear – this isn't a fluke, it was deliberate. Jonson used exactly the same ploy in his later play T-H-E A-L-C-H-E-M-I-S-T.

Audiences would not have noticed this word trickery unless they read a printed version of the play, which was published the following year, so it's clear that Jonson took an interest in people reading his work as well as watching it.

As far as we know, however, Shakespeare didn't play this acrostic game. Why not? One reason is that Shakespeare seems to have been less interested in getting his plays published than Jonson was. It is often said of Shakespeare that he wrote his plays to be performed, not to be read (which hasn't stopped school pupils having to spend countless hours doing just that), so there would have been little motivation to insert acrostics into his plays.

As we have seen, the only works of his where he was likely to have been involved with the printing process were his two narrative poems, *Venus and Adonis* and *The Rape of Lucrece*. Both were published early in his writing career, and there are no obvious acrostics in them. However, that hasn't stopped numerologists looking for them elsewhere. Since Shakespeare wasn't interested in printed versions of his plays, if there are acrostics and other hidden messages to be found, these must have been the work of the printers. And this is where the conspiracy theories begin.

One of Shakespeare's contemporaries was Sir Francis Bacon, a philosopher, scientist and statesman. As a young man, Bacon spent time in France where he studied statecraft and, as part of it, learned the important diplomatic craft of how to send secret messages, otherwise known as cryptography. Intrigued by this subject, Bacon devised his own ciphering system (using 'cipher' in its modern sense of code-making, rather than to mean 'zero'). His idea was to hide messages within regular text.

Bacon's ingenious binary system resembled the way that computers communicate using only the digits 0 and 1. In Bacon's case, messages would be hidden using slightly different printing typefaces, in which (say) a bold letter would represent a 1 and plain would be a zero. It meant that any innocent-looking sentence could be used to convey a hidden message, by breaking up text into groups of five letters. Using this code, the cheerful-sounding statement 'The **Quee**n be faring **well** to**day**' would actually carry the more grim message 'DYING'.[44]

Bacon explained that, using his system, anything could be used to signify anything (*'omnia par omnia'*). Little did he realize what a can of worms he was opening, because there would be many who would take this literally and use Shakespeare's text to find any messages that they wanted to find.

Since the late eighteenth century, there have been those who question whether the plays that are accredited to William Shakespeare were actually written by him. The main justification seems to be a snobbish one: how could somebody born in a Warwickshire backwater who didn't attend university become such a learned master playwright?

If somebody else had written the plays, they had done a good job of keeping their involvement secret. What literate person might be capable of such secrecy? Why, Sir Francis Bacon of course.

In the 1850s, an American woman by the name of Delia Bacon (no relation) became convinced that her namesake was the true author of Shakespeare's plays, and her investigations began a movement that continues to this day of people who identify themselves as 'Baconians'. To support their theory, Baconians have looked for secret messages embedded in Shakespeare's plays. And there are plenty of them, if you spend long enough looking for them.

In Act 1, Scene 2 of *The Tempest*, in the First Folio edition of Shakespeare's works, is this innocuous looking passage.

> For thou muſt now know farther.
> *Mira.* You haue often
> Begun to tell me what I am, but ſtopt
> And left me to a booteleſſe Inquiſition,
> Concluding, ſtay, not yet.

---

44 Note the plain and bold text. Using 00000 = A, 00001 = B etc, 'The **Qu**' is 00011 (D), '**een** be' is 11000 (Y) and so on. See the appendix for a fuller explanation of Bacon's cipher system.

I've added an L-shaped black line on the left. Follow the capital letters down and then across. and you see F BACon.

Need more evidence? In *Love's Labour's Lost* Shakespeare plants the word *Honorificabilitudinitatibus*. At 27 letters, it is the longest word in all of his work. What's it doing there? Does it contain a hidden message? In 1906, Isaac Hull Platt revealed that it is an anagram of the Latin phrase *Hi ludi, F. Baconis nati, tuiti orbi*, which translates roughly as: 'These plays, born of F Bacon, saved for the world'. And, no doubt, there are plenty of messages that can be found by applying Bacon's own ciphering system to Shakespeare's text.

Francis Bacon is not, however, the only contender for the title of true author of Shakespeare's work. Another faction, the Oxfordians, believe that Edward de Vere, the Earl of Oxford, wrote the plays. Their case is somewhat undermined by the fact that de Vere died in 1604, before several of Shakespeare's greatest plays were writtten, but the Oxfordians don't let this minor inconvenience get in the way of a good conspiracy. They too have found mentions of their man hidden within Shakespeare's work. One of their favourite pieces of evidence comes, ironically, from the engraved epitaph beneath the monument of Shakespeare himself, in Stratford-upon-Avon. The first sentence of the epitaph is in Latin, and the first eight words, written in pairs, look like this:

IVDICIO PYLIVM,[45]
GENIO SOCRATEM,
ARTE MARONEM
TERRA TEGIT

---

45 Meaning 'A Pylian in judgement, a Socrates in genius, a Maro [Virgil] in art, the earth buries him'.

Reading vertically, the second letters spell the word VERE. Was the engraver recording the identity of the true Shakespeare? And (to avoid being too obvious) were they sneakily choosing the second letter of every other word? Hmm.

These discoveries might feel like more than coincidences, but that is all that they are. Once again, you can get anything to mean anything if you look hard enough. There are some who believe that Ben Jonson wrote some of Shakespeare's plays. The American science fiction author John Sladek found a (tongue in cheek) anagram of that 'Baconian' word *Honorificabilitudinitatibus* which suggests that it was actually Ben Jonson who wrote a 'lifted batch' of Shakespeare's works. As the anagram says:

I B. Ionsonii, uurit a lift'd batch[46]

My favourite example is an anagram of the most famous Shakespearean passage of all. Cory Calhoun discovered that

To be or not to be: that is the question, whether tis nobler in the mind to suffer the slings and arrows of outrageous fortune.

is an anagram of:

In one of the Bard's best-thought-of tragedies, our insistent hero, Hamlet, queries on two fronts about how life turns rotten.

Did Shakespeare use his famous line to ingeniously embed a summary of the entire plot of the play? Or is this just a demonstration of what hokum this all is?[47]

---

46 The letters I and J were interchangeable, and printers often wrote 'w's as 'uu'.
47 On the subject of anagrams, 'William Shakespeare' is an anagram of 'I am a weakish speller', 'I'll make a wise phrase' and 'Hear me as I will speak'. Read into that what you will.

NUMEROLOGY AND CODES

How many numbers is in nouns?

THE MERRY WIVES OF WINDSOR

There is one other way in which some believe messages were hidden within Shakespeare, and that is through numbers. Nobody seriously argues that 'three score and ten' means anything other than 70. Curiously, however, the fact that 'three score and ten' famously features in the King James Bible has led some to believe that Shakespeare had a hand in writing the translation, as a favour to the king. As supporting evidence, it has been noted that in Psalm 46, the 46th word is 'Shake' and the 46th word from the end is 'Speare'. Add to this the fact that Shakespeare was 46 in the year when the King James Bible was printed and the conspiracy is complete. (Let's quietly ignore the fact that Shakespeare had turned 47 by the time the Bible was actually published, and that the counting of 46 words from the end of Psalm 46 conveniently ignores the word 'Selah' which is a sort of 'Amen' at the end of the psalm.)

Rather than looking at the numbers mentioned in Shakespeare's text, numerologists prefer to look at the letters. If you assign a number to each letter of the alphabet, with a = 1, b = 2 and so on, you can add up the values of letters in each word to create a number, and use these to find hidden messages.

For this to work, Baconians require the alphabet to have only 24 letters (with J and U omitted, as I and V were typically used in their place). Now take the word BACON. Its value is 2 + 1 + 3 + 14 + 13 = 33. The value of FRANCIS is 67. His full name therefore adds to 33 + 67 = 100, which feels numerically interesting, although it isn't clear what it means. Still, Baconians felt 100 must be significant. And it didn't escape their notice that 33 and 67 are one third and two thirds of 100 (rounded to the nearest whole number). Baconians put considerable effort into finding examples of words that add to Francis, Bacon or both.

One interesting example comes on page 56 of the Histories section of the First Folio, in Act 4 of *Henry IV Part 1*. One of the characters is Francis and his name appears copiously:

Prince:	Come hither Francis
Francis:	My Lord
Prince:	How long hast thou to serve, Francis?
Francis:	Forsooth five years and as much as to . . .
Poines:	Francis
Francis:	Anon, anon sir

And on it goes. A few lines further down the prince says:

Anon Francis? No Francis, but tomorrow Francis, or Francis on Thursday, or indeed Francis when thou wilt. But Francis.

That's six Francis's in a single quotation! It feels over the top and unnatural to repeat the name so often. What's going on? The mystery deepens when you count up the total number of times that Francis appears on the page. It is exactly 33, the Bacon number we calculated earlier.

Has somebody deliberately added gratuitous mentions of Francis to create this mystical number, as a message to tell us that Francis Bacon was involved in this work? Was it Bacon himself who was working with the printer?

Three pages earlier (on page 53 of that section of the First Folio) a character announces he has a Gammon of Bacon (with a capital B). This is one of only four mentions of Bacon in the whole of Shakespeare. One of the others is in *The Merry Wives of Windsor* and appears on a different page 53, in the Comedies section, and again, the B is capitalized. Is 53 significant?[48] There was a popular spiritual movement at this time known as Rosicrucianism, a Masonic-like brotherhood that operated in the shadows and promised to reveal the secrets of the physical and spiritual world. Some have claimed that Bacon was a follower of this movement, and they have found tenuous ways in which 53 is linked with Rosicrucianism.

But, with all these things, you have to ask *why* anyone would go to all this trouble to hide these messages? And if they did want to convey that Francis Bacon (or somebody else) was involved in Shakespeare's work, wouldn't they have found a neater and more systematic way of doing it?

The mysticism of numerology has even extended beyond Shakespeare's works to the design of the Globe theatre itself. In a popular 2007 episode of *Dr Who* entitled *The Shakespeare Code*, the Doctor observes that the theatre is a 14-sided tetradecagon, the same as the number of lines in a sonnet. This is historically inaccurate – experts believe that the Globe was actually a 20-sided icosagon. However, the nearby Rose theatre did indeed have 14 sides. Was there some profound reason to connect the shape of the theatre with a form of poetry? Probably not.

At least one serious and influential mathematician in Shakespeare's world would have taken all of this quite seriously – our old friend John Dee. His name was perfect for somebody who would take an interest in numerology. JOHN has four letters (the number of the four elements – earth, wind, fire and water), and DEE has three letters (representing the holy trinity of the Father, Son and Holy Spirit).

And what do you get if you add 4 + 3? Why, 7 of course. Dee was only too aware of that number's significance.

---

48 The consensus is that Shakespeare died on his birthday, 23 April 1616. He was entering his 53rd year on the day of his death, and that number appears on his monument. Baconians and Oxfordians have found tortuous ways to use this fact to back up their case.

SHAKESPEARE'S CALCULATOR

What's in a name?
ROMEO AND JULIET

I first set out on this quest to find links between Shakespeare and math after noticing his playfulness with numbers. Since then, I've discovered that Shakespeare lived in a world where new mathematical ideas were emerging in every walk of life, and that he was not only aware of them, but he often embraced them, too.

But was Shakespeare a *mathematician*? No. Only a few learned people of his time would have been given that description. However, Shakespeare could probably be described as an 'arithmetician'. This word has fallen out of use, but in Shakespeare's day arithmeticians were essential in a world of burgeoning trade, where accurate calculations of weights and money using fiddly units were a daily necessity. Shakespeare certainly displays many of the attributes of an arithmetician in the confident way in which he handles numbers.

Shakespeare uses the word arithmetician just once in his plays. In *Othello*, Iago is talking dismissively about his second-in-command:

And what was he?
Forsooth, a great arithmetician,
One Michael Cassio, a Florentine,
A fellow almost damn'd in a fair wife;
That never set a squadron in the field,
Nor the division of a battle knows
More than a spinster; unless the bookish theoric

Iago starts off sounding complimentary – calling him a 'great arithmetician' – but it goes downhill from there. 'A fellow . . . that never set a squadron in the field nor the division of a battle knows'. This man might be good on theory, but he has no practical expertise at all. (It's a description that is often used about pure mathematicians today.)

And here we find one final mathematical connection in Shakespeare's work.

He is Cassio.

Recognize that name? If you've ever wondered about that electronic calculator that is the most popular brand in the UK, and where it got its name ...

... then you'll be disappointed to learn that the calculator is actually nothing to do with Shakespeare. 'Casio' (with one 's') is a Japanese brand named after its founder, Tadao Kashio. As befits this final chapter on spurious connections, this link between Casio and Cassio, Shakespeare and mathematics, is nothing more than a coincidence.

And yet, if he were around today, Shakespeare would probably have enjoyed the pun.

FINIS.

APPENDIX:
MATHEMATICAL ASIDES

TACTICS FOR WINNING AT
THREE MEN'S MORRIS

If you have first move in Three Men's Morris and you are allowed to place your piece in the centre, you can always guarantee a win. Suppose you are playing white, and your opponent (black) places their piece in one of the corners:

To win, your next piece should go a 'knight's move' away from black, i.e. one along and one diagonal. For example, you can go middle left. You now have two in a line, so black is forced to block you, middle right:

Black is now threatening a line down the right-hand side, so you have no choice but to put your third piece bottom right. That gives you two in a diagonal, so black is forced to put their third piece top left:

Now that all the pieces are on the board, you can move a piece to an adjacent junction. You have a choice, but why not move the centre piece down to the bottom centre:

Black would love to block your bottom line but can't do it in a single move. Wherever they move (e.g. middle right to centre), you win by moving middle left to bottom left:

There's a similar strategy if black starts by placing their piece on the middle of one of the edges. And that is why the first player in Three Men's Morris is not allowed to go to the centre with their first move.

With perfect play, the game never ends and becomes a stalemate, but in practice somebody almost always makes a mistake that enables their opponent to win.

THE PYTHAGOREAN OCTAVE
AND DORIAN MODE

The modern seven-note musical scale that is the basis of most music in Western cultures can be traced back to ancient Greece, but it took hundreds of years to evolve into the notes we recognize today. The idea was to have pairs of notes that sounded good together. I mentioned in Chapter 6 that two strings, metal bars or hollow tubes whose lengths are in the ratio 2:1 (the 'octave') and 3:2 (in modern music this is slightly confusingly known as a perfect fifth) sound particularly pleasing together. Other whole-number ratios also sound good but, as a rule, the smaller the numbers in the ratio of two notes, the better they sound together. For example, two notes in the ratio 5:3 will sound better than notes in the ratio 43:29.

It's possible to build a scale of notes that (mostly) sound good together entirely from the ratios 2:1 and 3:2. One way to do this is to start with the base note, which we can give a frequency of 1 unit, and then use the ratio 3:2 to go up three notes and down three notes from there. Going up, you get 3/2, and then continue to multiply by 3/2, so next comes 9/4, and then 27/8, each note one and a half times the

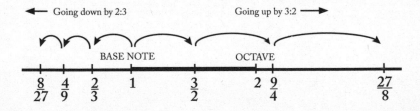

frequency of the one below, which ensures they sound good together. Going down you keep multiplying by 2/3, to get 2/3, 4/9 and 8/27.

However, to make a scale, we want all of the notes to be inside the range 1 (the bottom note) to 2 (the top note). In the scale we just created, only one of the notes (3/2) is in that range. Fortunately we can move the other notes into the range by doubling or halving their lengths – this won't change the 'letter' of any note, it'll just move it an octave higher or lower. After repeated halving and doubling we end up with six notes in between the bottom and top notes, making eight notes in all (hence 'octave'):

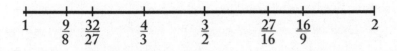

The fractions look quite evenly spaced along the number line, but notice that 9/8 and 32/27 are close together, and so are 27/16 and 16/9. These smaller gaps are known as semitones (or half steps), and the other gaps are known as tones (or whole steps).

If you were to divide strings into these fractions and pluck them, you would get a scale that sounds very close to what you'd hear when playing the white notes on a piano. But it would sound like the white notes starting at 'D', not the notes starting at 'C'. This is the Dorian mode mentioned on page 104, and it's partly because this is arguably the simplest music scale to build that the Dorian mode was the most popular mode for tunes in medieval Europe.

SHAKESPEARE, ORANGE AND SIR ISAAC NEWTON

The line from *King John* about adding 'another hue unto the rainbow' seems oddly prescient, because within a hundred years of Shakespeare's death another hue would indeed be added to the rainbow. In fact, there would be another two.

Sir Isaac Newton is best known for his work on gravity, but his second great achievement was his all-encompassing research into light, published in his second magnum opus, *Optics*.

Much of Newton's work was (unbeknown to him) simply replicating the findings of Thomas Harriot the previous century, but among the new revelations was his groundbreaking theory that light behaves as both a wave and a particle. However, he had a less scientifically rigorous theory on the colours of the rainbow.

Newton observed that, although the colours are spread across a spectrum, the colours at the two ends of the spectrum, red and violet, could be combined to make a new red-violet colour (what we would now call magenta). This led Newton to think that the colours of a rainbow should actually be set out in a circle rather than in a line, and he presented them in a diagram. This was the first definitive setting-out of rainbow colours since Aristotle, and now there were seven colours instead of three.

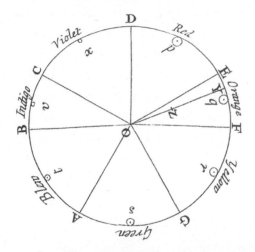

Newton's circular diagram showing the colours of the rainbow.

The curious thing, however, is that Newton did not observe these seven colours himself. He admitted to only really being able to see five colours – red, yellow, green, blue and violet (having bumped purple off his list of favoured colour names). He claimed that it was a friend who had identified the other two colours for him.

And that is how orange got promoted to being one of the rainbow colours. But Shakespeare should be given some credit for popularizing it as a colour name in the first place.

The other colour to gain promotion to the rainbow – this time less deserved – was indigo. You have to use a lot of imagination to see a band of indigo in a rainbow.[49] Much more obvious to the naked eye is a band of light blue, what we would call 'cyan', but because Newton ignored it, or just lumped it into the category called 'blue', cyan is these days firmly in the second division of colour names. Today we tend to call it 'light blue' (except in printer ink cartridges).

---

49 Modern representations of rainbow or spectrum often show only six colours. Indigo is ignored in the LGBTQ+ Pride Flag, the logo for Apple computers, and also in the prism image on the cover of Pink Floyd's album *Dark Side of the Moon*.

Whether his 'friend' could see seven colours or not, the fact is that Newton *wanted* there to be seven colours. Despite his scientific rigour, Newton was also something of a mystic. The ancient Pythagorean belief that everything in heaven and earth is connected in harmony resonated with him, and seven colours would fit perfectly into this pattern.

Take another look at Newton's colour circle. To the left of each colour name there is a letter. Starting with the first colour of the rainbow, the letters go as follows:

D   Red
E   Orange
F   Yellow
G   Green
A   Blue
B   Indigo
C   Violet

And having gone around the circle you are now back to D, where you started.

Does this cycle of letters remind you of anything? These are the seven notes of the musical scale. Newton was convinced that the blending and harmony of colours was directly analogous to the blending of notes to make harmonious chords. In the same way that playing the notes D and A together on a piano keyboard make a pleasant, harmonious sound, Newton felt that colours could be combined in a harmonious way.

For once, Newton was wrong. Blending colours doesn't work in the same way as blending musical notes. But in deciding that the order of musical colours begins at D rather than the more obvious A, or even C (which is these days the central musical note), Newton was inadvertently giving a nod to the Dorian mode of music that we met in Chapter 6. Rainbows were the visible manifestation of the music of the spheres.

PHILIP HENSLOWE'S DIARY

No records remain of the accounts and operation of the Globe theatre, but we can get a good understanding of how theatres in Southwark operated thanks to the journal of Philip Henslowe, who managed the Rose theatre, a neighbouring theatre to the Globe, and the first such establishment on the south bank of the Thames.

Henslowe kept a diary from 1592 until 1609, which almost perfectly spanned Shakespeare's career even though the two had little involvement with each other. The company based at Henslowe's theatre, the Admiral's Men, were staunch rivals of Shakespeare's company, the Chamberlain's (and later King's) Men, though they were known to collaborate, for example during times of plague.

The bulk of the diary entries detail Henslowe's expenses, loans and box office receipts, but various notes on other matters are dotted throughout, as if Henslowe sometimes needed a spare bit of paper to jot down a reminder of some pressing matter.

It's Henslowe's diary that tells us that a playwright would typically be paid £5 or £6 for a new play (the equivalent of a few thousand pounds in today's money), and that new plays would sometimes bring in receipts close to that amount just on their first performance. *Henry VI* made nearly £4 on its first night.

Henslowe's bookkeeping is far from perfect. He makes numerous errors when adding up figures, and in his dates (there are a couple of mentions of 31 June, for example). What's particularly striking is the way he jumps from Roman to Arabic numerals on a whim. Much of the time he writes dates in Arabic numerals but sums in Roman. For example, the first line of the ledger below says 'Received for 10 May 1596 ------------ XXXs' i.e. 30 shillings. It is thought these are his personal takings for that's night performance. There would have been money for the other shareholders and actors too.

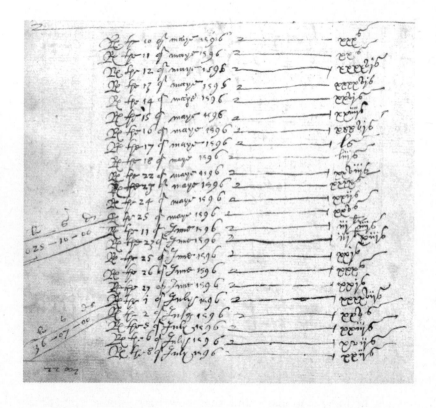

On another page he squeezes in some addition. The columns are headed pounds, 'li' (which is short for the word livres), shillings (s) and pence (d) – remember there were 20 shillings in a pound. Note how Henslowe can't resist putting dots onto two of the 1s to make them Roman i's.

li	s	d
211	9	0
188	11	6
400	0	6

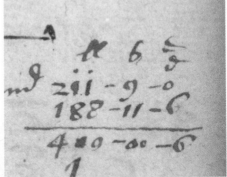

And in 1597 all his entries used Arabic numerals:

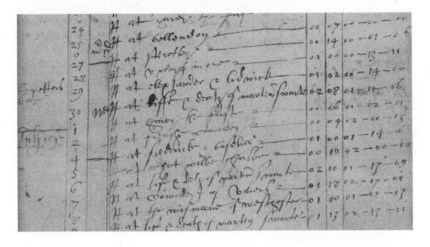

FRANCIS BACON'S CIPHER SYSTEM

Francis Bacon's ingenious cipher system allowed people to embed secret messages in any text. It's possible that he never actually used it in practice; he merely presented it as an interesting thought experiment. By using two different fonts, such as bold and plain, the letters of the alphabet were encrypted as follows:

Code	Converted to binary numbers	Letter
aaaaa	00000	A
aaaab	00001	B
aaaba	00010	C
aaabb	00011	D
aabaa	00100	E
aabab	00101	F
aabba	00110	G
aabbb	00111	H
abaaa	01000	I,J
abaab	01001	K
ababa	01010	L
ababb	01011	M
abbaa	01100	N
abbab	01101	O
abbba	01110	P
abbbb	01111	Q

baaaa	10000	R
baaab	10001	S
baaba	10010	T
baabb	10011	U, V
babaa	10100	W
babab	10101	X
babba	10110	Y
babbb	10111	Z

In the early 1900s, William and Elizabeth Friedman were inspired by the Bacon cipher to become expert cryptanalysts. Their work in cryptology and deciphering Japanese messages was a valuable part of the USA war effort.

TIMELINES

SHAKESPEARE'S LIFE AND HISTORIC EVENTS

Shakespeare's Life	Year	Historic Events
	1558	Elizabeth becomes queen
Shakespeare is born in Stratford-upon-Avon	1564	Galileo is born in Pisa Elizabeth sets up Company of Mines Royal
	1568	Mercator's map projection introduced
	1577	Sighting of supernova in the skies
	1580	Francis Drake completes circumnavigation of the world
Shakespeare marries Anne Hathaway	1582	Gregorian Calendar introduced
Birth of Shakespeare's children, Susanna followed by twins Hamnet and Judith	1583–85	
Shakespeare's 'lost years', almost nothing known about his life or activities.	1585	Walter Raleigh oversees Roanoke settlement in America (within five years it vanished)
	1587	Mary Queen of Scots executed
	1588	Defeat of Spanish Armada
Shakespeare's first play performed (now known as *Henry VI Part 2*)	1591?	
Purchase of New Place, a large house in Stratford	1597	
Opening of the Globe theatre	1599	
	1600	Founding of the East India Company
Shakespeare's acting company become known as 'The King's Men'	1603	Death of Elizabeth I, coronation of King James I of England
	1605	Gunpowder plot is foiled
First performance of *Macbeth*	1606	
	1610	Galileo sees Jupiter's moons
	1611	King James Bible published
Final play that is attributed to Shakespeare, *Henry VIII*	1613	
Shakespeare dies in Stratford	1616	
Publication of the First Folio, a complete collection of Shakespeare's plays	1623	

SHAKESPEARE'S PLAYS AND POEMS

There is some dispute over when some of Shakespeare's plays and poems were written, and different sources provide different timelines. The list of plays below comes from the Royal Shakespeare Company, and some of the dates are deliberately vague. Other sources generally agree with this list, plus or minus a year. The poems known to have been by Shakespeare have been inserted into the timeline in italics.

Before 1592	The Taming of the Shrew
1591	Henry VI Part II
1591	Henry VI Part III
1590s	The Two Gentlemen of Verona
1591–92	Titus Andronicus
1592	Henry VI Part I
1592	Richard III
1593	*Venus and Adonis*
1594	*The Rape of Lucrece*
1594	The Comedy of Errors
1595–96	Love's Labour's Lost
1595–96	A Midsummer Night's Dream
1595–96	Romeo and Juliet
1595–96	Richard II
1595–97	King John
1596–7	The Merchant of Venice

1596–97	Henry IV Part I
1597–98	Henry IV Part II
1597–1601	The Merry Wives of Windsor
1598	Much Ado About Nothing
1599	Henry V
1599	As You Like It
1599	Julius Caesar
1600	Hamlet
1601	Twelfth Night
1601–02	Troilus and Cressida
1604	Othello
1604	Measure for Measure
1603–06	All's Well That Ends Well
1604–06	Timon of Athens
1605–06	King Lear
1606	Macbeth
1606–07	Antony and Cleopatra
1608	Coriolanus
1608	Pericles
1609	*The Sonnets***
1610	Cymbeline
1611	The Winter's Tale
1611	The Tempest
1613	Henry VIII*
1613–14	The Two Noble Kinsmen*

* Cowritten with John Fletcher.

** Probably written between 1590 and 1605, they were published in full in 1609.

BIBLIOGRAPHY

Many books have been helpful in building a picture of life in Shakespeare's world. I read all of the following cover to cover:

Bearman, Robert, *Shakespeare's Money*, Oxford University Press, 2016

Bryson, Bill, *Shakespeare: The World as a Stage*, Harper, 2016

Gamini, Salgado, *The Elizabethan Underworld*, Sutton, 1999

Kaufman, Miranda, *Black Tudors: The Untold Story*, Oneworld, 2018

MacGregor, Neil, *Shakespeare's Restless World*, Viking, 2013

Mortimer, Ian, *The Time Traveller's Guide to Tudor England*, Viking 2013

Morton, Giles, *Nathaniel's Nutmeg*, John Murray, 2000

Norton, Elizabeth, *The Lives of Tudor Women*, Head of Zeus, 2017

O'Farrell, Maggie, *Hamnet*, Tinder Press, 2020

Shapiro, James, *1599: A Year in the Life of William Shakespeare*, Faber, 2016

Smith, Emma, *The Making of Shakespeare's First Folio*, 2nd edition, Bodleian Library, 2023

Strathern, Paul, *The Other Renaissance: From Copernicus to Shakespeare*, Atlantic, 2023

Woolley, Benjamin, *The Queen's Conjuror: The Life and Magic of Dr Dee*, Flamingo, 2002

Important reference sources

Baynes, Thomas S., 'What Shakespeare learnt at school', in *Shakespeare Studies, and Essay on English Dictionaries*, Longmans, Green and Co., 1894

Chappell, D. Havens, *Shakespeare's Astronomy*, Astronomical Society of the Pacific, 1945

Chrisomalis, Stephen, 'Numbering by the books: the transition from Roman to Arabic numerals in the early English printing tradition', Academia, 2009, downloadable pdf

Daybell, James, *The Material Letter in Early Modern England: Manuscript Letters and the Culture and Practices of Letter-Writing, 1512–1635*, Palgrave, 2012

Dobson, M., Wells, S., Sharpe, W. and Sullivan, E., *The Oxford Companion to Shakespeare*, 2nd edition, Oxford University Press, 2015

Friedman, William F. and Friedman, Elizabeth S., *The Shakespearean Ciphers Examined*, Cambridge University Press, 1956 (illustrated edition, 2011)

Forsyth, Mark, *The Elements of Eloquence*, Icon Books, 2016

Howson, A.G., *A History of Mathematics Education in England*, Cambridge University Press, 1982

Hunt, Katherine, 'Jangling bells inside and outside the playhouse', in B. Barclay and D. Lindley (eds), *Shakespeare, Music and Performance*, pp. 71–83, Cambridge University Press, 2017

Kojoyan, Ani, 'Inter-textual relations between Reginald Scot's "The Discoverie of Witchcraft" and Shakespeare's "Macbeth"', *Armenian Folia Anglistika*, 2013

Kollerstrom, Nicholas, *The Secrets of the Seven Metals: A Bridge Between Heaven and Earth*, Lulu.com, 2015

Monroe Stowe, A., *English Grammar Schools in the Reign of Queen Elizabeth*, Leopold Classic Library, 2015

Rouse Ball, W.W., *A History of the Study of Mathematics at Cambridge*, Legare Street Press, 2022

Olson, D.W., Olson, M.S. and Doescher, R.L., 'The stars of Hamlet: Shakespeare's astronomical inspiration?', Free Library, 1998

Scot, Reginald, *The Discoverie of Witchcraft*, Dover, 2000 (reprint of sixteenth-century classic)

Tijms, Stephen, *Chance, Logic and Intuition*, World Scientific, 2021

Tyler, Ian, *Seathwaite Wad and the Mines of the Borrowdale Valley*, Blue Rock, 1995

ACKNOWLEDGMENTS

I would like to thank Chris Head, who has been an invaluable sounding board during this entire project. Chris also introduced me to Rebecca Macmillan, a Shakespeare improviser, who has brought a whole new perspective to Shakespeare and math that would never have occurred to me otherwise.

During my research, I've found myself consulting experts on a hugely diverse range of topics, and I've been overwhelmed by how generous they've been in sharing their knowledge. An early port of call was the Shakespeare Institute in Stratford-upon-Avon, where I was directed to Dr Simon Smith. Simon has been immensely helpful and supportive both in sharing his own expertise and in pointing me to others. I'm also grateful to have been able to tap into the expertise of Michael Dobson, the Director of that Institute. Other Shakespeare experts who were extremely helpful in answering my quirky questions about the playwright and his times include Professor Emma Smith, Patrick Spottiswoode, Professor John Drakakis, Dr Bob Bearman and Dr Ben Higgins.

It has also been a joy to be able to contact historians and archivists in some quite niche subject areas. My thanks, for example, to Morag Loader and Richard Geldard of the History of Tax Group at the Worshipful Company of Tax Advisers. I'm so glad you guys exist! And talking of Worshipful Companies, thanks also to Ruth Frendo and Mark Butler of, respectively, the Stationers and the Glovers.

Calista Lucy, Peter Jolly and Freddie Witts were very welcoming in showing me the original Henslowe diary at Dulwich College, as was Lindsey Armstrong at Shakespeare's Schoolroom in Stratford. I also had valuable input from Sarah Wheale (Head of Rare Books at Oxford's Weston Library), Martin Allen (Numismatist at Fitzwilliam Museum in Cambridge), Christina Faraday (Art Historian at Gonville and Caius College, Cambridge), and David Sherren (Cartography expert at Portsmouth Library).

Thank you also to Robyn Arianrhod for filling me in on Thomas Harriot, Ben Woolley on John Dee, David Parlett on the rules of Tudor games, Mark Hatton on the Seathwaite mines, Scott Blaber on real tennis, and Helen Arney on Tudor musical instruments. Add to that list Matthew Scroggs, Richard Harris, Michael Haslam for some great snippets – and sorry to anyone that I've left out.

As a source of reference, I had three website tabs open almost permanently. I'm grateful to the many people behind Open Source Shakespeare, to LitCharts for their Modern English Shakespeare Translations, and to Ben Crystal and David Crystal for Shakespeare's Words, all of them incredibly useful online resources.

I had encouraging feedback on the first draft from Timandra Harkness and Elinor Flavell when I needed it most. My wife Elaine and daughter Jenna were spot on with their comments on the second and Mairi Sutherland and Colin Beveridge on the third.

It goes without saying that any errors or omissions are down to me.

Thank you to my friend Andrew Jeffrey, whose enthusiastic collaboration in our math workshop in Stratford-upon-Avon provided the first spark for this project.

Finally, my thanks to Erika and the rest the team at Atlantic Books, and in particular to my editor Ed Faulkner, who, over a beer at the Great Exhibition pub, uttered the immortal line 'you do realize there's a book in this, don't you?'

IMAGE CREDITS

INDEX

Note: Page numbers in *italics* indicate illustrations.

ABOUT THE AUTHOR

Rob Eastaway has authored and coauthored several books that connect math with everyday life, including the bestsellers *Why Do Buses Come in Threes?* and *Maths on the Back of an Envelope*. Since 2004, he has been the director of Maths Inspiration, an interactive lecture series encouraging higher math education that has reached over 250,000 teenagers in the United Kingdom, New York and Sydney. For five years, he was also the editor of *New Scientist* magazine's weekly math puzzle. He has over 30 years' experience presenting on math to audiences of all ages, including as a keynote speaker at educational conferences and as a regular guest on BBC Radio. In 2017, he received the Zeeman medal for excellence in the public communication of mathematics, jointly awarded by the London Mathematical Society and the Institute of Maths and its Applications. He lives in London, roughly two-score-and-eight furlongs from Shakespeare's historic Globe Theatre.

robeastaway.com | @robeastaway